Physical Properties of Materials for Engineers

Volume I

Author

Daniel D. Pollock
Professor of Engineering
State University of New York at Buffalo
Buffalo, New York

CRC Press, Inc.
Boca Raton, Florida

Library of Congress Cataloging in Publication Data

Pollock, Daniel D.
Physical properties of materials for engineers.

Includes bibliographies and indexes.
1. Solids. 2. Materials. I. Title.
QC176.P64 620.1'12 81-839
ISBN 0-8493-6200-8 (set) AACR2
ISBN 0-8493-6201-6 (v. 1)
ISBN 0-8493-6202-4 (v. 2)
ISBN 0-8493-6203-2 (v. 3)

Direct all inquiries to CRC Press, Inc., 2000 Corporate Blvd., N.W., Boca Raton, Florida, 33431.

Second Printing, 1984

International Standard Book Number 0-8493-6200-8 (Complete Set)
International Standard Book Number 0-8493-6201-6 (Volume I)
International Standard Book Number 0-8493-6202-4 (Volume II)
International Standard Book Number 0-8493-6203-2 (Volume III)

Library of Congress Card Number 81-839
Printed in the United States

PREFACE

Many new materials and devices which were designed to possess specific properties for special purposes have become available in the recent past. These have had their origins in basic scientific concepts. Engineers must understand the bases for these developments so that they can make optimum use of available materials and further advance the existing technology as new materials appear. The main objective of this text is to provide engineers and engineering students a unified, elementary treatment of the basic physical relationships governing those properties of materials of greatest interest and utility.

Many texts on solid state physics, written primarily for advanced undergraduate physics courses, make use of sophisticated mathematical derivations in which only the most significant parts are given; the intermediate steps are left to the reader to provide. This makes it difficult for the average engineer to follow and has the effect of discouraging or "turning off" many readers. Other texts are not much more than surveys of "materials science" and provide little insight into the nature of the phenomena.

This text represents an attempt to provide a middle ground between these extremes. It is designed to explain the origin and nature of the most widely used physical properties of materials to engineers; thus, it prepares them to understand and to utilize materials more effectively. It also may be used as a textbook for senior undergraduate and first-year graduate students.

Practicing engineers will find this text helpful in getting up to date. Readers with some familiarity with this field will be able to follow the presentations with ease. Engineering students and those taking physics courses will find this book to be a useful source of examples of applications of the theory to commercially available materials as well as for uncomplicated explanations of physical properties. In many cases alternate explanations have been provided for clarity.

An effort has been made to keep the mathematics as unsophisticated as possible without "watering down" or distorting the concepts. In practically all cases only a mastery of elementary calculus is required to follow the derivations. All of the "algebra" is shown and no steps in the derivations are considered to be obvious to the reader. Explanations are provided in cases where more advanced mathematics is employed The problems have been designed to promote understanding rather than mathematical agility or computational skill.

The introductory chapters are intended to span the gap between the classical mechanics, which is familiar to engineers and engineering students, and the quantum mechanics, which usually is unfamiliar. The limitations of the classical approach are shown in elementary ways and the need for the quantum mechanics is demonstrated. The quantum mechanics is developed directly from this by the use of uncomplicated examples of various phenomena. The degree to which the quantum mechanics is presented is sufficient for the understanding of the physical properties discussed in the subsequent chapters; it also provides a sound basis for more advanced study.

Introductory sections are given which guide the reader to the topic under consideration. The basic physical relationships are provided. These are drawn from concepts and properties which are known to those with engineering backgrounds; they lead the reader into the topic of interest. In some cases small amounts of material are repeated for the sake of clarity and convenience. Some topics, frequently presented as separate chapters in physics texts, have been incorporated in various sections in which they are directly applicable to materials. Lattice dynamics is one of the subjects treated in this way. Where appropriate, sections covering the properties of commercially available materials are included and discussed. This approach provides more comprehensive presentations which can be readily followed and applied by the reader.

Since this text is intended for readers with engineering backgrounds, some of the topics often presented in solid state physics books have been omitted. The reader is, however, provided with a suitable foundation upon which to pursue such topics elsewhere. The fundamentals of solid state physics are indispensable to the understanding of the properties of materials; these have been retained. Thus, the approach and content of this text are unique in that they include the properties and applications as well as the theory of those major types of real materials which are most frequently employed by engineers. This is rarely, if ever, done in current physics texts.

On the other hand, important subjects marginally included, or omitted, from many physics texts have been incorporated. Chapter 6 (Electrical Resistivities and Temperature Coefficients of Metals and Alloys) and Chapter 7 (Thermoelectric Properties of Metals and Alloys) are good examples of this. These chapters are unique in that similar material does not appear in any text of which I am aware. Sections of these chapters include the basic physical theories and their relationships to phase equilibria as well as their application to the design of alloys with special sets of electrical properties and to the explanations of the properties of commercially available alloys. The very wide use of these types of alloys makes it necessary that engineers thoroughly understand the mechanisms responsible for their optimum applications and their limitations. Other topics of primary importance to engineers which are normally included in solid state physics texts also have been incorporated.

The background required for this text includes elementary calculus, first-year, college-level physics and chemistry, and one course in physical metallurgy or materials science. Information required beyond these levels has been incorporated where needed. This makes it possible to accommodate the needs of readers where there is a wide range of background and capability; it also permits self-study.

The first five chapters introduce, explain, and develop the modern theory of solids; these are considered to constitute the minimum basis for any text of this type. Various other sections, or chapters, may then be studied, depending upon the interests of the reader and the emphasis desired. One combination of topics could be selected by electrical engineers, another set by metallurgical engineers, still another group by mechanical engineers, etc. Courses in materials engineering could be organized in similar ways. It should be noted, however, that all of the major topics included in this text represent physical properties employed by, and of significance to, most engineers at some time during their careers.

Note should be made that the units used in each of the topics are those currently employed by engineers working with materials in that area. The use of a single system of units would be counterproductive. Means for conversions to other units are given in the text for convenience and in the appendix.

I wish to express my deep appreciation to two of my former teachers for the insights and approaches to solid state phenomena which they provided early in my career. Professor C. W. Curtis, of Lehigh University, and Dr. F. E. Jaumot, then associated with the University of Pennsylvania, have been continuing sources of inspiration. In addition, some of the illustrations given in Dr. Curtis' lectures have served, with his permission, as models for the equivalents given here. Similarly, I am indebted to Dr. Jaumot for permission to use his clear approaches to reciprocal space, Brillouin zone theory, and the elementary theory of alloy phases as a basis for those used here.

I am deeply grateful to the American Society for Testing and Materials for permission to condense the contents and to use the illustrations from the monograph, *The Theory and Properties of Thermocouple Elements,* STP492, 1971, written by the author. This material is presented in Chapter 7 (Volume II).

Acknowledgment is also made of the assistance provided by Mr. James Stewart for his cooperation and assistance in the preparation of the illustrations.

I am very grateful to Donna George for her unfailing patience and help in typing the manuscript.

Credits are given with the individual tables and figures.

Daniel D. Pollock

PHYSICAL PROPERTIES OF MATERIALS FOR ENGINEERS

Daniel D. Pollock

Volume I

Beginnings of Quantum Mechanics
Waves and Particles
The Schrödinger Wave Equation
Thermal Properties of Nonconductors
Classification of Solids

Volume II

Electrical Resistivities and Temperature Coefficients of Metals and Alloys
Thermoelectric Properties of Metals and Alloys
Diamagnetic and Paramagnetic Effects
Ferromagnetism

Volume III

Physical Factors in Phase Formation
Semiconductors
Dielectric Properties
Useful Physical Constants
Conversion of Units

TABLE OF CONTENTS

Volume I

Chapter 1

BEGINNINGS OF QUANTUM MECHANICS

The purpose of this chapter is to provide the basis for an insight into the nature of quantum effects. It is not intended to describe the selected physical phenomena in detail. Such information is readily available in numerous physics texts. It is intended to show how these physical reactions provide the beginnings for quantum mechanics as ultimately applied to real materials. The historical approach is employed as a means for the gradual introduction of the fundamental concepts.

The Newtonian mechanics are based upon the ideas that the variables of a system, such as energy or momentum, can be precisely known at any given time or position. This approach leads to the descriptions of the behaviors of systems on a macroscopic scale. Newton was able to describe geometric optics on this basis by considering that a beam of light consists of a stream of discrete corpuscles.

The Newtonian, or classical, approach cannot explain diffraction phenomena. The wave theory of the nature of light proposed by Huyghens was able to account for these phenomena as well as for geometric optics. Maxwell later showed that light comprised a small portion of the electromagnetic wave spectrum. Thus, the wave-particle question developed. It was to play an important part in the modern quantum mechanic theory.

1.1. BLACK-BODY RADIATION

All solids and some liquids glow when sufficiently heated (above about 700°C); they give off radiation. At low temperatures the intensity of this radiation is too weak to be seen. The atoms in these solids and liquids are very close together so that they are not independent oscillators. This interdependence (see Chapter 4) and interaction results in a continuous spectrum of wavelengths. The radiation from low-pressure gases, where each atom or molecule behaves nearly independently, is discrete and occurs only at the wavelengths of the spectral lines of the gases. Of immediate interest is the radiation behavior of solids. This is shown schematically in Figure 1-1.

This spectrum is approximately the same for all solids at a given temperature and approaches what is known as the black-body spectrum. A perfect black body is one which absorbs all incident radiation (zero reflectivity). Therefore, all radiation emanating from a black body originates within it.

The classical attempt at an explanation for this phenomenon is based upon the equipartition of the average energy of each mode of vibration of a group of independent oscillators, each with its own frequency. This leads to an expression for the radiant energy per unit volume, $E(\lambda)$, given by

$$E(\lambda) = \frac{8\pi k_B T}{\lambda^4} \tag{1-1}$$

in which k_B is Boltzmann's constant, T is the absolute temperature and λ is the wavelength. This cannot explain the curves given in Figure 1-1. At very long wavelengths, however, Equation 1-1 approaches agreement with the experimental data.

It will be noted in Figure 1-1 that the wavelengths corresponding to the maxima, λ_M, vary inversely with increasing temperature. This may be stated as

$$\lambda_M T = \text{const} \tag{1-2}$$

and is known as the Wien displacement law. It is one of the bases of optical pyrometry.

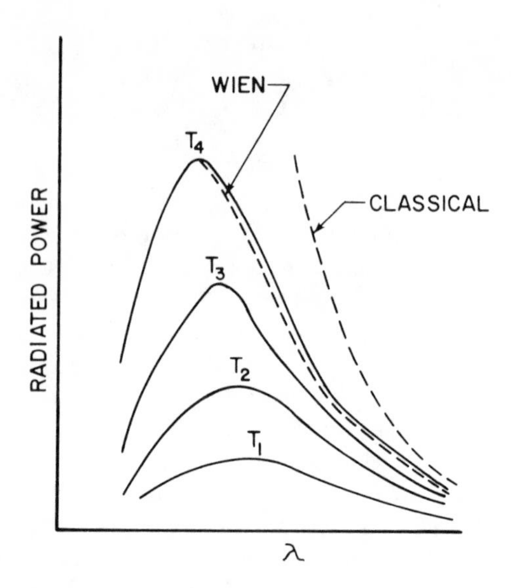

FIGURE 1-1. Radiated power as a function of wavelength for several temperatures: $T_1 < T_2 < T_3 < T_4$. (Modified from Richtmyer, F. K., Kennard, E. H., and Lauritsen, T., *Introduction to Modern Physics*, 5th ed., McGraw-Hill, New York, 1955, 131. With permission.)

This had a long history of use in the estimation of temperatures by the color of a heated piece of metal, long before physics became a science.

When the radiated energies are compared at different temperatures, the maximum energy is found to vary as T^5, or

$$E_{max} = g\left(\lambda_M, T\right) T^5 \tag{1-3}$$

or, since λ and T are inversely related,

$$E_{max} = \frac{h(\lambda, T)}{\lambda_M^5} \tag{1-4}$$

Wien suggested that this behavior could be given more closely by

$$E_{max} = \frac{C_1}{\lambda_M^5} \cdot \frac{1}{e^{C_2/\lambda T}} \tag{1-5}$$

The constants C_1 and C_2 were empirical and not defined. This relationship accurately fits the observed data from the region of the maximum to the shorter wavelengths. It does not fit the experimental data from the maximum to the longer wavelengths. Despite this shortcoming, Wien's work constituted a significant advance.

1.2. PLANCK'S LAW

Planck was not only able to overcome the difficulties encountered by his predecessors, but included their findings as special cases of his analysis. This was achieved by an entirely new approach to the problem.

Planck (1901) considered that the radiating surface was composed of electric dipole oscillators, and that there should be a relationship between the energy of the radiation and the frequency of the oscillator. Implicit in all of the prior work was the idea that the energy of an oscillator could vary in a continuous way. Planck's major contribution lies in assuming that this is not so. His assumptions are as follows:

1. The energy of an oscillator is of the form

$$E = nh\nu \tag{1-6}$$

   where n is an integer, h is the constant of proportionality, now known as Planck's constant, ($h = 6.626 \times 10^{-27}$ erg − sec) and ν is the frequency of the oscillation. This means that the energy of the oscillator can no longer be continuous, but can only take on discrete values determined by the integer n.
2. Since the energy of an oscillator cannot be continuous, its transitions from one energy level to another must be associated with the absorption or emission of energy.
3. The energy gained or lost in such transitions must be in discrete, or quantum, amounts.

These assumptions constituted a great departure from the classical approach and, as such, became one of the bases of modern physics. They were to have far-reaching consequences.

When the number of vibrational modes which can be assumed by an oscillator is taken into account along with the average energy of each mode, it is found that

$$E(\lambda, T)d\lambda = \frac{8\pi d\lambda}{\lambda^4} \cdot \frac{h\nu}{e^{h\nu/k_BT} - 1}$$

where k_B is Boltzmann's constant. Or, recalling that $\lambda\nu = c$ (c = the speed of light), this becomes

$$E(\lambda, T) = \frac{8\pi ch}{\lambda^5} \cdot \frac{1}{e^{ch/\lambda k_BT} - 1} \tag{1-7}$$

This general law gives a very good fit with the observed data over a wide range of wavelengths. It relates the energy associated with any temperature and wavelength, not just with that of the wavelength of the maximum energy. For small values of λ and T, $\exp ch/\lambda k_BT \gg 1$, so Equation 1-7 becomes the same as Equation 1-5 and the constants are now defined as $C_1 = 8\pi ch$ and $C_2 = ch/k_B$. For large values of λ, the denominator of Equation 1-7 may be approximated by a series approximation as follows:

$$\exp ch/\lambda k_BT - 1 \approx 1 + ch/\lambda k_BT + + + - 1 = ch/\lambda k_BT$$

Equation 1-7 now becomes

$$E(\lambda, T) = \frac{8\pi ch}{\lambda^5} \cdot \frac{\lambda k_BT}{ch} = \frac{8\pi k_BT}{\lambda^4} \tag{1-8}$$

Equation 1-8 is the same as Equation 1-1 which was found by classical means. Thus,

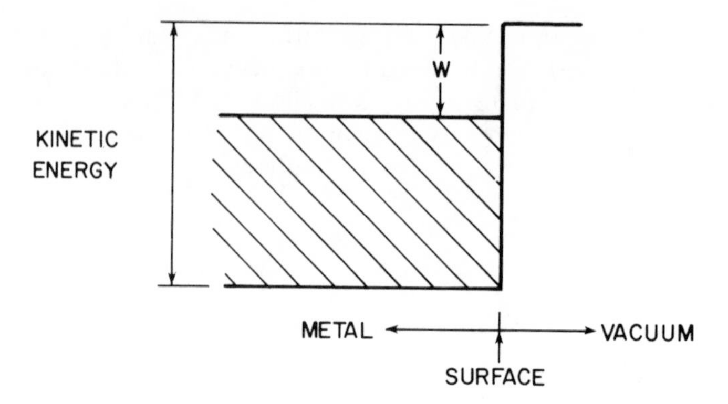

FIGURE 1-2. Work function, W, and energy barrier at the surface of a metal.

for suitably large values of λ the quantum expression reduces to the classical expression.

The main point of this work is that the assumption of a discrete energy behavior, rather than a continuous one, gives rise to an expression for black-body radiation which agrees with the observed behavior. It could not have been obtained on the basis of a continuous energy spectrum. The idea of discrete rather than continuous behavior was used to explain other physical effects and became one of the basic concepts of quantum mechanics.

1.3. PHOTOELECTRIC EFFECT

Hertz (1887) found that ultraviolet light from one electric spark made it easier (required a lower voltage) for sparks to jump across a gap between two electrodes in another adjacent circuit. It is now known that the electrons emitted from the surface of a material, when it is irradiated by light of suitable wavelength, are responsible for this. When a voltage (potential difference) exists between a pair of electrodes, the radiation-generated electrons emitted by one electrode will flow to the other electrode. Lower voltages are required to strike an arc between the electrodes when such externally induced electron flows take place between them. The behavior observed by Hertz is a manifestation of the photoelectric effect.

For the sake of this discussion it is assumed that the electron emitter is a metal. Electrons do not normally leave the surface of a metal unless sufficient energy is added. This may be in the form of radiant energy, as previously noted, or as thermal energy (thermionic emission). Thus, the metal volume may be considered as constituting a lower energy environment for the freely moving electrons than that external to itself. If this were not the case, electrons would not require the additional energy for their removal from the metal. This situation may be pictured as shown schematically in Figure 1-2. According to this figure, the most energetic electron within the metal would require a minimum increase in energy equal to W in order to leave the metal. W is known as the work function and is a constant for a given, clean metal surface; it is different for each metal. Also see Figure 3-4(b) in Chapter 3.

According to classical theory, the electric field associated with the radiation should excite, or remove, the electron, if it is strong enough. Thus, if the kinetic energy increment which the radiation imparts to the electron is equal to or greater than W, the electron will be emitted. It was thought that the greater the intensity of the incident radiation, the greater would be the kinetic energy of the emitted electrons, and, since

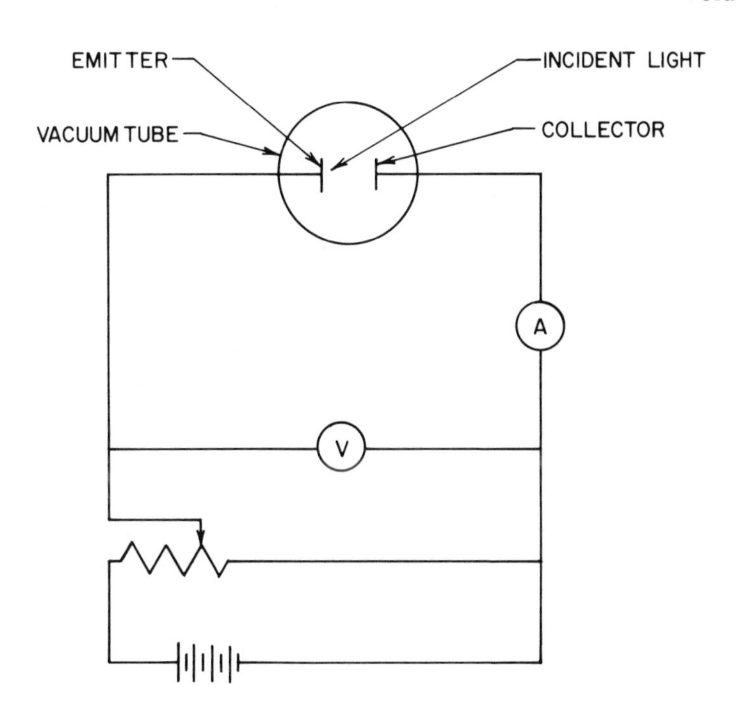

FIGURE 1-3. Schematic diagram of photoelectric apparatus. (Modified from Sproull, R. L., *Modern Physics,* John Wiley & Sons, New York, 1956, 76. With permission.)

W is the minimum excitation energy, a lower intensity limit would thus exist below which no electrons would be emitted. On this basis, the kinetic energy of the emitted electrons was expected to be a function only of the intensity of the radiation. As will be shown, this was in error. The kinetic energy of the emitted electrons is actually a function of the frequency of the incident radiation.

Observations of the photoelectric effect can be made with apparatus similar to that shown schematically in Figure 1-3. This apparatus permits the application of either a retarding voltage or an accelerating voltage to be applied between the electrodes in the vacuum tube. The retarding voltage, V, can be adjusted so that $eV = K.E. \geqslant W$, where e is the charge on an electron and K.E. is its kinetic energy as a result of irradiation. In other words, the retarding voltage can be made to be such that the electrical energy between the electrodes would just balance the kinetic energy of the irradiated electrons and no current would flow between the electrodes. This provides a direct measure of the kinetic energy imparted to the electrons by the radiation.

Using monochromatic incident light directed upon the emitter, and varying its intensity, the results shown in Figure 1-4 are observed. As the negative-retarding voltage is increased from zero, fewer electrons can reach the collector. A voltage is reached, V_o, such that no electrons can leave the emitter. This gives the maximum kinetic energy as eV_o. As the intensity of the monochromatic light is increased, V_o remains constant. The number of emitted electrons is increased, increasing the current, but their kinetic energy remains constant (V_o is constant); it is not a function of intensity. This means that, contrary to the classical theory, the intensity of the incident radiation has no effect upon the energy of the electrons.

When the frequency of the incident light is increased and its intensity is held constant, the behavior shown in Figure 1-5 is observed. The higher the frequency of the radiation, the greater $V_{o,i}$ becomes. This means that the energy of the emitted electrons

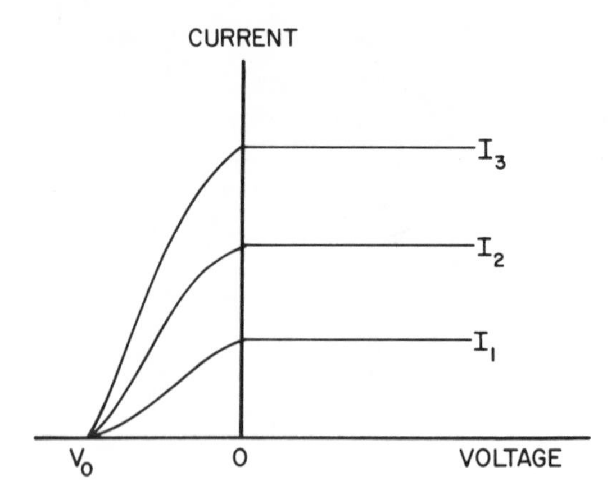

FIGURE 1-4. Variation of electric current with the intensity of monochromatic radiation as a function of voltage: $I_3 > I_2 > I_1$. (After Sproull, R. L., *Modern Physics*, John Wiley & Sons, New York, 1956, 77. With permission.)

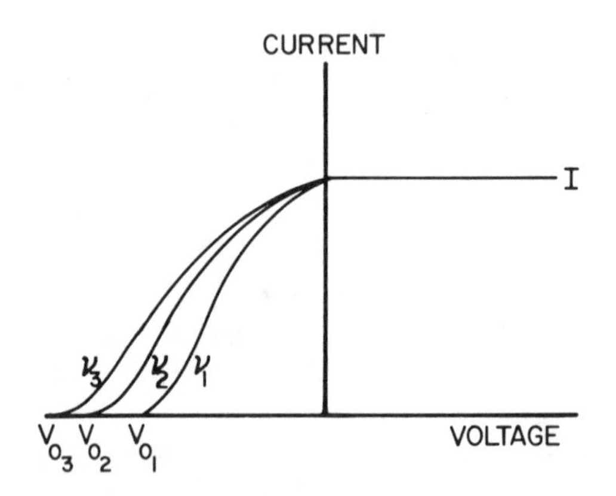

FIGURE 1-5. Variation of retarding potential with constant intensity monochromatic light of different frequencies. Note that the currents resulting from increasing frequencies actually lie in a narrow band in the first quadrant. (Modified from Sproull, R. L., *Modern Physics*, John Wiley & Sons, New York, 1956, 78. With permission.)

is increasing and that greater retarding voltages are required to counteract their flow. In other words, increasing the frequency of the incident radiation produces more energetic electrons as is shown by the greater retarding voltages. This experiment clearly demonstrates that the frequency of the incident radiation, not its intensity, determines the energy of the emitted electrons.

A plot of the retarding voltages vs. frequency is shown in Figure 1-6. This is a linear relationship where the voltage intercept is w. The product of $w \cdot e = W$, the work function. Similar plots for other metals show that all are linear and have identical slopes, but that each has a different work function. The classical theory could not have predicted this.

Einstein (1905) explained this behavior by making assumptions similar to those of Planck. He quantized the radiation as Planck had quantized the oscillators. The assumptions used to explain photoelectric behavior are:

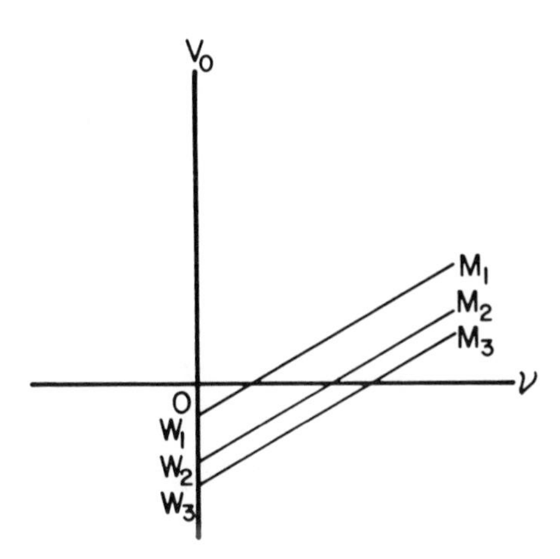

FIGURE 1-6. Retarding voltage as a function of the frequency of irradiation for three metals. (After Sproull, R. L., *Modern Physics*, John Wiley & Sons, New York, 1956, 78. With permission.)

1. The incident radiation is made up of discrete pulses called photons.
2. These photons are absorbed or given off in discrete amounts.
3. Each photon has an energy which is an integral multiple of $h\nu$.
4. The photons behave like waves of corresponding frequency.

The fourth assumption is an important contribution since it implies duality and anticipates de Broglie (Chapter 2).

Now consider an electron with energy W, Figure 1-2, when it absorbs energy from an incident photon. The additional energy absorbed by it will be equal to $h\nu$. Assume that the emitted electron leaves along a path perpendicular to the emitting surface. It will require a minimum energy of W to leave the given metallic surface. The net energy of the emerging electron will be

$$\text{K.E.} = eV = h\nu - W \tag{1-9}$$

When this expression is differentiated with respect to ν it gives, since W is a constant for a given metal,

$$\frac{dV}{d\nu} = \frac{h}{e} \tag{1-10}$$

Equation 1-10 predicts a constant slope which is independent of the emitting source, in agreement with the observed data. Thus, the slope is the same for any metal, the work function being a different constant depending upon the given metal. This theoretical approach constitutes another important milestone in the understanding of physical behavior. It explains the photoelectric effect in terms of quantized radiation. This could not have been explained by a classical theory which is based upon a continuous radiation spectrum.

1.4. SPECTROGRAPHIC BACKGROUND

According to classical theory, a continuous spectrum is expected when matter is

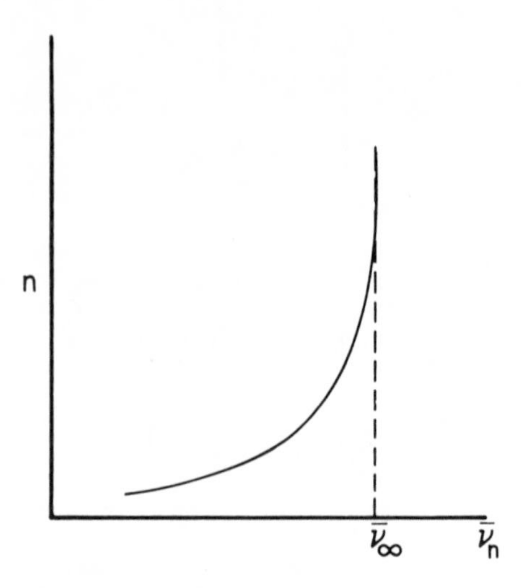

FIGURE 1-7. Schematic representation of Rydberg's findings.

excited by an arc, or spark, as in spectrography. It is found experimentally that only "lines", or discrete radiations occur when the light is diffracted. These lines occur at wavelengths which are characteristic of the element(s) being examined. This is one of the bases for chemical analysis by spectrographic means.

Balmer (1885) first showed the relationship between observed spectral lines for hydrogen by the empirical formula

$$\lambda = \text{const}\ \frac{n^2}{n^2 - 4} \tag{1-11}$$

where n is an integer equal to or greater than 3. He posed the question whether this relationship could be applied to the spectra of other elements.

In investigating this question, Rydberg (1890) found that different series of spectral lines associated with elements could be described by

$$\bar{\nu}_n = \bar{\nu}_\infty - \frac{R_y}{(n + \mu)^2} \tag{1-12}$$

where $\bar{\nu}_n$ is the reciprocal of the wavelength of the line, $\bar{\nu}_\infty$ is the upper limit of the series, R_y is a constant now called the Rydberg, n is an integer and μ is a constant. This behavior is shown schematically in Figure 1-7. The parameters $\bar{\nu}_\infty$ and μ are different constants for each of the four general series of spectral lines. These series were named by the early workers in spectra as s = sharp, p = principal, d = diffuse, and f = fundamental. These designations are now also used to define the second quantum number, ℓ, of the electron states of an atom.

Equation 1-11 is a special case of Equation 1-12. Rydberg also gave a general relationship which, for spectra of the hydrogen atom, can be reduced to

$$\bar{\nu}_n = R_y \left[\frac{1}{n_1^2} - \frac{1}{n_2^2}\right] \tag{1-13}$$

Equation 1-13 gives the series of lines for hydrogen when the values for the integers n_i given in Table 1-1 are used.

Table 1-1
VALUES OF INTEGERS REQUIRED TO GIVE THE SPECTRAL SERIES OF HYDROGEN

Series	n_1	n_2	Radiation
Lyman	1	2,3,4,...	Ultraviolet
Balmer	2	3,4,5,...	Visible
Paschen	3	4,5,6,...	Near infrared
Brackett	4	5,6,7,...	Far infrared
Pfund	5	6,7,8,...	Far infrared

Note: It will be noted that all of the variable quantities in Equations 1-11, 1-12, and 1-13 are integers. This would lead to the idea that discrete behavior, rather than continuous behavior, must be involved in the structure of the atom itself. This was shown to be the case in Bohr's attempt to explain spectral lines in terms of the electron structure of atoms.

1.5. THE BOHR HYDROGEN ATOM

The Bohr approach is based upon Rutherford's model of the atom. This model pictures the atom as consisting of a positive nucleus surrounded in an unspecified way by a number of electrons equal to the nuclear charge to maintain electrical neutrality. An important change in the earlier model was that Bohr assumed that the electrons were contained in discrete circular orbits and held there by coulombic forces without emitting radiant energy. This meant that their energies had to be discrete. Thus, Bohr was forced to assume that their angular momenta in these orbits was in integral multiples of $h/2\pi$. The frequency of a spectral line was then assumed to be proportional to the difference in energy between two such orbits. This idea is essentially the same as the discrete energy assumptions made by Planck and Einstein.

Based upon the assumption of coulombic attraction, the force between the nucleus and an electron is

$$F = \frac{eE}{r^2} \qquad (1\text{-}14)$$

where e is the charge on the electron, E is the charge on the nucleus and r is the distance between them. The charge on the nucleus is given by

$$E = eZ \qquad (1\text{-}15)$$

in which Z is the atomic number. Upon the substitution of Equation 1-15 into Equation 14, it becomes

$$F = \frac{e^2 Z}{r^2} \qquad (1\text{-}16)$$

The centripetal force is mv^2/r, where m is the mass and v is the velocity of the electron. The centripetal and coulombic forces acting upon the electron are equated to give, for an electron in a given orbit,

$$F = \frac{mv^2}{r} = \frac{e^2 Z}{r^2} \qquad (1\text{-}17)$$

or

$$v^2 = \frac{e^2 Z}{mr} \qquad (1\text{-}18)$$

This permits the calculation of the kinetic energy, K.E., from

$$K.E. = \frac{1}{2} mv^2 = \frac{1}{2} m \cdot \frac{e^2 Z}{mr} = \frac{e^2 Z}{2r} \qquad (1\text{-}19)$$

If the potential energy, P.E., of the electron is taken as zero when it is infinitely far from the nucleus, then the

$$P.E. = -\frac{e^2 Z}{r} = -2\ K.E. \qquad (1\text{-}20)$$

The total energy is E_T = K.E. + P.E. Using Equation 1-20, the total energy is found to be

$$E_T = -\frac{e^2 Z}{2r} \qquad (1\text{-}21)$$

In terms of angular frequency, ω, and frequency, ν, where $\omega = 2\pi\nu$,

$$K.E. = \frac{1}{2} mr^2\omega^2 = \frac{1}{2} mr^2 (2\pi\nu)^2 \qquad (1\text{-}22)$$

Upon differentiation with respect to ν, the change in energy is given by

$$\Delta E = 4\pi^2 m\, r^2 \nu\, d\nu \qquad (1\text{-}23)$$

Bohr's new ideas contained the assumption that the electronic orbits were discrete and that an energy-frequency proportionality existed, in the same way as had Planck and Einstein, or

$$\Delta E = nh\nu \qquad (1\text{-}24)$$

When Equations 1-23 and 1-24 are equated

$$nh\nu = 4\pi^2 mr^2 \nu d\nu$$

or

$$\frac{nh}{2\pi} = 2\pi mr^2 d\nu \qquad (1\text{-}25)$$

The angular momentum, $p(\theta)$, is given by

$$p(\theta) = mr^2\omega \qquad (1\text{-}26)$$

Upon differentiation this becomes

$$\Delta p(\theta) = mr^2\Delta\omega \quad (1\text{-}27)$$

and since

$$\Delta\omega = 2\pi\Delta\nu$$

Equation 1-27 becomes

$$\Delta P(\theta) = 2\pi mr^2\Delta\nu \quad (1\text{-}28)$$

By comparison of Equations 1-25 and 1-28

$$\Delta p(\theta) = \frac{nh}{2\pi} \quad (1\text{-}29)$$

This is the basis for the original Bohr assumption that the angular momenta of the electrons had to be in multiples of $h/2\pi$; it quantizes the angular momenta and guarantees discrete orbits. It also provides that the electrons will occupy circular, nonradiating orbits which will not "run down" into the nucleus and give continuous spectra. This is discussed further in Section 1.7.

Now by means of Equation 1-26, noting that $\omega = v/r$,

$$p(\theta) = mr^2\frac{v}{r}$$

or

$$mv = \frac{p(\theta)}{r}$$

But momentum is quantized, not continuous, so from Equation 1-29 this becomes

$$mv = \frac{nh}{2\pi r} \quad (1\text{-}30)$$

Equation 1-30 will be used to obtain an expression for the Bohr radius.

Starting with an expression based upon Equation 1-17 in which both sides are multiplied by the mass, m,

$$\frac{m^2v^2}{r} = \frac{m\,e^2Z}{r^2}$$

or

$$(mv)^2 = \frac{m\,e^2Z}{r} \quad (1\text{-}31)$$

The square of Equation 1-30 is equated to Equation 1-31 and the resulting expression is solved for r:

$$\frac{n^2h^2}{4\pi^2r^2} = \frac{m\,e^2Z}{r}$$

$$r = \frac{n^2h^2}{4\pi^2me^2Z} \quad (1\text{-}32)$$

This is the expression for the Bohr radius. It may also be expressed as

$$r = \text{const}\,\frac{n^2}{Z} \tag{1-33}$$

where the constant is 0.528×10^{-8} cm. For the case of hydrogen in its lowest, or ground, state n and Z equal unity and r = 0.528 Å. This is sometimes used as a unit of length in describing electronic structures and probability densities.

The kinetic energy of an electron may be obtained by the substitution of Equation 1-32 into Equation 1-19:

$$\text{K.E.} = \frac{e^2 Z}{2r} = \frac{e^2 Z}{2} \cdot \frac{4\pi^2 m e^2 Z}{n^2 h^2}$$

or, simplifying and using Equation 1-21

$$\text{K.E.} = 2\pi^2 m \left[\frac{e^2 Z}{nh}\right]^2 = -E_T \tag{1-34}$$

Now, on the basis of the assumption that the frequency of a given spectral line is proportional to the difference between two energy states, Equation 1-34 can be used in the following way:

$$h\nu = E_{T_2} - E_{T_1} = \text{K.E.}_1 - \text{K.E.}_2$$

or

$$h\nu = 2\pi^2 m \left[\frac{e^2 Z}{h}\right]^2 \cdot \left[\frac{1}{n_1^2} - \frac{1}{n_2^2}\right] \tag{1-35}$$

or

$$\nu = \frac{2\pi^2 m e^4 Z^2}{h^3}\left[\frac{1}{n_1^2} - \frac{1}{n_2^2}\right] \tag{1-36}$$

Since the behavior of the nucleus was not taken into account by Bohr in this derivation, Equation 1-36 does not include small deviations in frequency arising from nuclear motions.

Equation 1-36 looks very much like Equation 1-13. The expression which was derived empirically by Rydberg can be obtained from Equation 1-36 by noting that $\bar{\nu}_n = 1/\lambda = \nu/c$, or

$$\bar{\nu}_n = \frac{1}{\lambda} = \frac{2\pi^2 m e^4 Z^2}{h^3 c}\left[\frac{1}{n_1^2} - \frac{1}{n_2^2}\right] \tag{1-37}$$

All of the factors in Equation 1-37 are constant for a given element except the integers n_i. Thus, a value may be obtained for the Rydberg constant:

$$R_y = \frac{2\pi^2 m e^4}{h^3 c} = 1.097 \times 10^5/\text{cm}$$

This is in excellent agreement with the value of R_y obtained experimentally from the

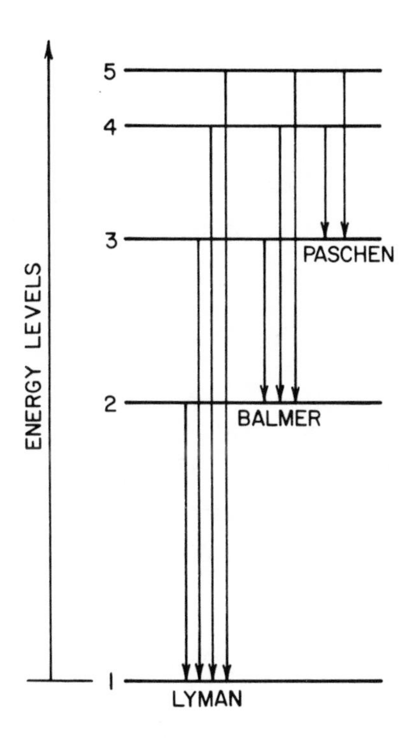

FIGURE 1-8. Schematic diagram for spectra of hydrogen.

Balmer lines of hydrogen. The value of R_y is not precisely constant, but increases very slightly with atomic number from H to Sc. It remains essentially constant beyond Sc ($Z > 21$).

The variable integers, n_i, are the principle quantum numbers. The substitution of appropriate integers into Equation 1-37 can account for all of the spectral lines of hydrogen (see Table 1-1). This is also shown graphically in Figure 1-8. It is clearly shown that radiation is only emitted when an electron drops to a lower energy level. Its excess energy is given off as radiation.

The expression for the Bohr radius, Equation 1-32, may be used to obtain an expression for the velocity of an electron from the kinetic energy, Equation 1-19:

$$\text{K.E.} = \tfrac{1}{2} mv^2 = \frac{e^2 Z}{2r} = \frac{e^2 Z}{2} \cdot \frac{4\pi^2 me^2 Z}{n^2 h^2}$$

Upon multiplication and extraction of the square root, the velocity of an electron is given by

$$v = \frac{2\pi e^2 Z}{nh} \qquad (1\text{-}38)$$

or,

$$v = \text{const}\, \frac{Z}{n}$$

Thus, for a given element, Z is constant; the velocity and energy of an electron would be expected to vary inversely with the principle quantum number. Electrons in outer shells would be expected to have smaller velocities and energies than those in inner levels in a given atom.

The relative velocity of an electron may be obtained from Equation 1-38 by dividing both of its sides by c, the velocity of light:

$$\frac{v}{c} = \frac{2\pi e^2}{ch} \cdot \frac{Z}{n} = \frac{1}{137.29} \cdot \frac{Z}{n} \qquad (1\text{-}38a)$$

In the case of the hydrogen atom in the ground state, Z and n equal unity; the velocity of the electron, as computed from this equation, is slightly less than 1% of the speed of light.

In addition to the description of the hydrogen atom in terms of its spectrum, Equation 1-37, its permissible electron states may be obtained with good accuracy from Equations 1-21 and 1-32 in terms of energy as

$$E_n = -\frac{e^2 z}{2r} = \frac{e^2 z}{2} \cdot \frac{4\pi^2 m e^2 Z}{n^2 h^2} = -2\pi^2 m \left[\frac{e^2 Z}{nh}\right]^2$$

$$= -\frac{13.6}{n^2}\ (\text{eV}) \qquad (1\text{-}34a)$$

This also gives the Rydberg, in another form, as 13.6 eV. Recent work on "high Rydberg" states on highly excited atoms, $n \sim 80$, shows them to be astonishingly different from those described in Table 1.1.

1.6. BOHR CORRESPONDENCE PRINCIPLE

Up to this point consideration has only been given to cases where the principle quantum numbers are small (see Table 1-1, e.g.). What happens to Bohr's equations for the condition in which n becomes large? Earlier discussion has described one analogous case. When the wavelength in Planck's equation, (Equation 1-7) becomes very long it reduces to Equation 1-8 which is identical to Equation 1-1. In other words, the quantum-mechanical relationship is transformed into the classical equation for long wavelengths.

Equation 1-36 will be used to answer this question for the case where the quantum numbers are large and represent adjacent energy levels. Equation 1-36 may be rewritten as

$$\nu = \frac{2\pi^2 m e^4 Z^2}{h^3}\left[\frac{1}{n_1^2} - \frac{1}{n_2^2}\right] = \frac{2\pi^2 m e^4 Z^2}{h^3}\left[\frac{n_2^2 - n_1^2}{n_1^2 n_2^2}\right] \qquad (1\text{-}39)$$

Use now is made of the conditions placed upon the quantum numbers. Since these are adjacent levels,

$$n_2 - n_1 = 1 \text{ and } n_2 = n_1 + 1$$

The substitution of these values for the quantity within the brackets of Equation 1-39 gives, after factoring the numerator,

$$\frac{n_2^2 - n_1^2}{n_2^2 n_1^2} = \frac{(n_2 - n_1)(n_2 + n_1)}{n_1^2 (n_1 + 1)^2} = \frac{1 \cdot [(n_1 + 1) + n_1]}{n_1^2 (n_1 + 1)^2}$$

$$= \frac{2n_1 + 1}{n_1^2 (n_1 + 1)^2}$$

The quantum number is large so that

$$n_1 = n >> 1$$

and

$$n_1 + 1 \simeq n$$

These relationships simplify the fraction given above:

$$\frac{n_2^2 - n_1^2}{n_2^2 n_1^2} \simeq \frac{2n}{n^4} = \frac{2}{n^3}$$

This result is substituted into Equation 1-39 to give

$$\nu \simeq \frac{2\pi^2 me^4 Z^2}{h^3} \cdot \frac{2}{n^3} = \frac{4\pi^2 me^4 Z^2}{h^3 n^3} \qquad (1\text{-}40)$$

Now consider the classical approach to frequency.

$$2\pi\nu = \omega = \frac{v}{r}$$

$$\nu = \frac{v}{2\pi r} \qquad (1\text{-}41)$$

Equations 1-32 and 1-38 provide expressions for the factors r and v, respectively. These are substituted into Equation 1-41 to give

$$\nu = \frac{1}{2\pi} \cdot \frac{2\pi e^2 Z}{nh} \cdot \frac{4\pi^2 me^2 Z}{n^2 h^2} = \frac{4\pi^2 me^4 Z^2}{n^3 h^3} \qquad (1\text{-}42)$$

Equation 1-42 is the same as Equation 1-40. Thus, as n was made sufficiently large, the Bohr equation for frequency was transformed into the classical equation. More generally, at the appropriate limit the quantum results must agree with the classical results. This qualification is called the correspondence principle. It constitutes a test for the validity of quantum-mechanical treatments, since under suitable conditions they must approach the classical cases. This constraint establishes the classical mechanics as a special case of the quantum mechanics.

1.7. COMMENTS ON THE BOHR MODEL

Bohr's revolutionary ideas provided a significant portion of the bases for the modern understanding of atomic structure. Their limitations will be appreciated when the ideas presented in Chapters 2 and 3 are understood.

The Bohr model successfully explained the spectra of hydrogen and included the findings of prior investigators. It also was able to explain the spectra of atoms with larger atomic numbers in a qualitative way.

The only "justification" for the assumption that the permissible electron orbits were those where the angular orbital momenta are multiples of $h/2\pi$ is given by Equation 1-29. This was necessary in order to explain the discrete spectral lines. If this limitation had not been made, continuous spectra (contrary to observation) would have been permissible. The necessity for this restriction lies in the Rutherford model of the atom. If no discrete orbits were required, an electron radiating energy must have its total energy decreased. If this proceeds in a continuous way, its radius about the nucleus (Equation 1-22) must also similarly decrease; a continuous spectrum would then be emitted. Ultimately, the electron would fall into the nucleus. The behavior postulated by Bohr, rather than that just described, agrees with the observed data. This is the reason for the postulation that the electrons were in discrete, nonradiating orbits.

The Bohr model also was able to account for the earlier spectrographic findings. Equation 1-37 accomplished this, at least for helium and lithium. It gave a more fundamental basis for the calculation of the Rydberg constant beyond that of empirical determination.

Bohr, Sommerfeld, and others tried unsuccessfully to improve this model, primarily by the use of elliptical orbits. This was not possible because other quantum-mechanical effects were not yet known.

1.8. PROBLEMS

1. Compare the calculated radiated energies as determined by the classical and Planck relationships at 1000 K, 1550 K and 1650 K for wavelengths of 1.5×10^4 Å, 3.0×10^4 Å, and 6.0×10^4 Å.
2. Calculate the minimum wavelength for photoemission from a metal whose work function is 3.0 eV.
3. Calculate the kinetic, potential, and total energies of an electron in a hydrogen atom in the ground state.
4. Determine the energies of electrons whose wavelengths are 5, 50, 500, and 5000 Å.
5. Compare the wavelengths and energies involved in the first three lines of Lyman and Balmer series.
6. Approximate the velocity of an electron in a hydrogen atom in the ground state in units of cm/sec.
7. Use the expression for R_y contained in Equation 1-37 to obtain its equivalent in terms of energy.

1.9. REFERENCES

1. **Sproull, R. L.,** *Modern Physics,* John Wiley & Sons, New York, 1956.
2. **Richtmyer, F. K., Kennard, E. H., and Lauritsen, T.,** *Introduction to Modern Physics,* 5th ed., McGraw-Hill, New York, 1955.
3. **White, H. E.,** *Introduction to Atomic Spectra,* McGraw-Hill, New York, 1934.
4. **Blanchard, C. H., Burnett, C. R., Stover, R. G., and Weber, R. L.,** *Introduction to Modern Physics,* Prentice-Hall, Englewood Cliffs, N.J., 1969.

Chapter 2

WAVES AND PARTICLES

The wave- and particle-like behavior of electrons will be considered. Einstein assumed this duality for radiation (Section 1.3 in Chapter 1). Experiments involving electrons and diffraction effects will be described to illustrate this behavior. These will show the need for the statistical treatment of electrons.

The classical physics and the findings of Einstein unmistakeably show that light possesses both wave and particle properties. If light is corpuscular in nature, how can diffraction phenomena be explained? This question could be resolved if it were assumed that the waves and particles were associated in some way. It has already been assumed by Einstein and others that photons, wave quanta, etc., have energies associated with their frequencies: $E = h\nu$. The nature of this wave-particle relationship and its consequences will be considered in this chapter.

2.1. ELECTRON OPTICS

The optical behaviors of electrons (similar to those used in electron microscopes) will be employed to show some of the properties of electrons significant to an understanding of their nature.

The responses of electrons in electric and magnetic fields are evidence for their existence as charged particles. They move in straight lines in electric and magnetic fields and their direction of motion changes when the strength of the electric fields changes, neglecting magnetic fields. Under these conditions, electron beams may be considered to be matter waves which obey the laws of geometric optics. They can be made to reflect and refract. Consider Figure 2-1, in which the refraction of electrons is simulated. Electrons from the source enter chamber 1, across which is the uniform potential V_1. They pass through the opening between the evacuated chambers and enter chamber 2. This chamber has another uniform potential $V_2 > V_1$ across it. The two different electric fields set up by these potentials impart different velocities to the electrons, v_{p1} and v_{p2}. The direction of motion of the electrons will change when they enter the second field. This is the only effect of the potentials which must be considered here. The velocity components of the electrons perpendicular to the applied fields, v_y, remain unchanged. Thus, for electrons treated as matter particles

$$v_y = v_{p_1} \sin \theta_1 = v_{p_2} \sin \theta_2$$

or

$$\frac{\sin \theta_1}{\sin \theta_2} = \frac{v_{p_2}}{v_{p_1}} \qquad (2\text{-}1)$$

However, if the electron beam is treated as being composed of waves, Snell's law can be used to describe the refraction. This gives

$$\frac{\sin \theta_1}{\sin \theta_2} = \frac{v_1}{v_2} \qquad (2\text{-}2)$$

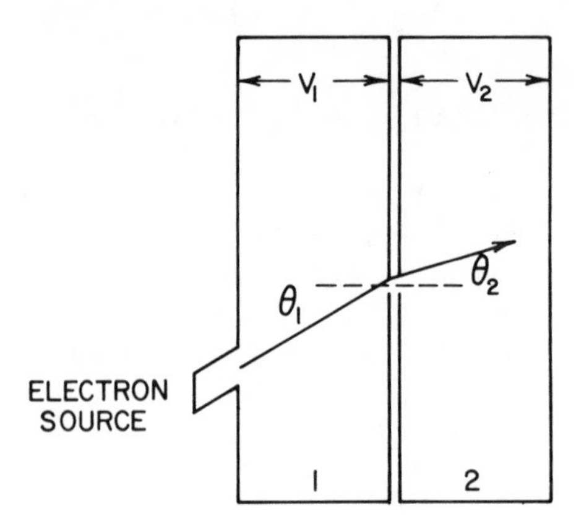

FIGURE 2-1. Schematic diagram showing refraction-like behavior of electrons. (After Richtmyer, F. K., Kennard, E. H., and Lauritsen, T., *Introduction to Modern Physics,* 5th ed., McGraw-Hill, New York, 1955, 176. With permission.)

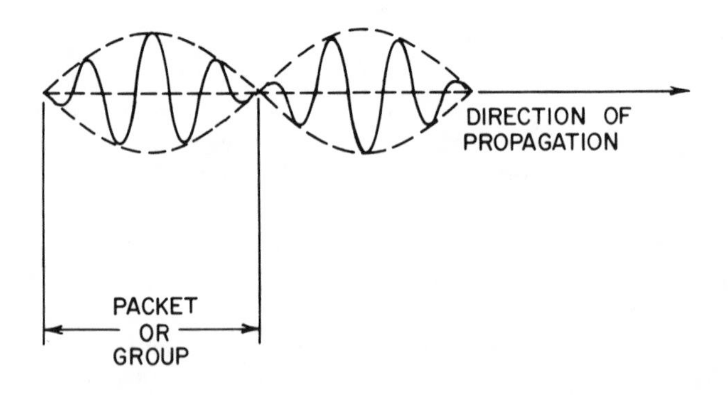

FIGURE 2-2. Schematic representation of wave packets.

a relationship which appears to contradict that given by Equation 2-1. This seeming contradiction arises from treating electrons as waves rather than as particles.

In order to resolve this question of apparent contradiction, it is necessary to consider the way in which waves and particles may be associated. The particle may be simulated by means of a group of waves, or a wave packet. This is shown in Figure 2-2.

If such a packet containing about 10^5 waves could represent a corpuscle of light, a stream of such corpuscles, or quanta, would give optical results in conformity with the classical geometric optics. However, since each packet may be constructed from the superposition of waves, this model might also account for the properties of light which can be explained by waves.

In a similar way, an electron may also be considered to be such a wave packet made of matter waves. In this case, an electron with an energy of one electron volt (1eV) would have a wavelength of 1.24×10^{-4} cm and a frequency of 2.4×10^{14} sec^{-1} associated with it.

Now consider the behavior of these packets. A packet will be observed to move in the direction of propagation with a group velocity v_g. The individual waves within the group may move with another velocity, their phase velocity, which is equal to v. Now assume that the waves have the properties of particles. Using Equation 1-6,

$$E = h\nu; \quad \nu = \frac{E}{h} \tag{2-3}$$

and substituting Einstein's relationship between energy and mass, m, for the energy, gives

$$\nu = \frac{mc^2}{h} \tag{2-4}$$

in which c is the velocity of light. This neglects relativistic changes in mass at high velocities. The moving particle, or packet, will have a momentum, p. Assume that the packet has a wavelength, λ, associated with it because of its energy-frequency relationship. Now assume further that this wavelength and the momentum of the particle are related to each other as

$$\lambda = \frac{h}{p} \tag{2-5}$$

This is the de Broglie equation and is discussed more fully in Section 2.2. The factor p is the momentum of a particle and it is equal to the product of its mass, m, and its velocity, v_p, so, upon making this substitution for p,

$$\lambda = \frac{h}{mv_p} \tag{2-6}$$

The wavelength and frequency are related to the velocity by

$$v = \lambda\nu \tag{2-7}$$

so that Equations 2-4 and 2-6 give, when substituted into Equation 2-7,

$$v = \frac{h}{mv_p} \cdot \frac{mc^2}{h} = \frac{c^2}{v_p} \tag{2-8}$$

Equation 2-8 shows that the velocity of the wave is inversely proportional to the velocity of the particle associated with it.

If the particle in question is an electron, Equation 2-8 resolves the previously noted "contradiction". Equation 2-1 was based upon the treatment of the electron as a particle; Equation 2-2 treated it as a wave. These treatments gave inverse results which are in agreement with Equation 2-8. The apparent "contradiction" is resolved.

Additional examination of the wave packet representing the electron (Figure 2-2), can further clarify its behavior. The waves within the packet can be represented by the superposition of two waves of slightly different wavelength. The larger amplitudes near the center of the packet then result from their constructive interference; the diminishing amplitudes on either side are caused by destructive interference. This leads to an expression for the group velocity as

$$v_g = v - \lambda \frac{dv}{d\lambda} = v - \frac{\lambda}{d\lambda} dv \qquad (2\text{-}9)$$

When applied to a wave packet representing an electron, this equation will shed further light upon the relationship between the group and phase velocities. Expressions must be obtained for dv and $\lambda/d\lambda$ to accomplish this.

To obtain an expression for the quantity dv, Equation 2-7 is used. Here a wave number, k, is defined as the reciprocal of the wavelength. When this substitution is made,

$$v = \lambda\nu; \quad \nu = \frac{v}{\lambda} = vk \qquad (2\text{-}10)$$

When this is differentiated and rearranged,

$$dv = \frac{d\nu - vdk}{k} \qquad (2\text{-}11)$$

The expression needed for $\lambda/d\lambda$ is obtained from the definition of the wave number given above. From this

$$k\lambda = 1$$

and upon differentiation this gives

$$\frac{\lambda}{d\lambda} = -\frac{k}{dk} \qquad (2\text{-}12)$$

The values given by Equations 2-11 and 2-12 are substituted into Equation 2-9 to obtain

$$v_g = v - \left(-\frac{k}{dk}\right)\left(\frac{d\nu - vdk}{k}\right)$$

The factor k vanishes and the group velocity is found to be

$$v_g = v + \frac{d\nu}{dk} - v = \frac{d\nu}{dk} \qquad (2\text{-}13)$$

Equation 2-3 is used in order to obtain greater insight into the relationship between v_g and v, since an expression for $d\nu$ may be obtained from it for use in Equation 2-13. This makes use of the relativistic expression for mass:

$$\nu = \frac{E}{h} = \frac{mc^2}{h} = \frac{m_o c^2}{h(1-\beta^2)^{1/2}} \tag{2-14}$$

where m_o is the rest mass of the particle and $\beta = v/c$. Upon differentiation, Equation 2-14 becomes

$$d\nu = \frac{m_o c^2}{h} - \frac{1}{2}(1-\beta^2)^{-3/2}(-2\beta)\,d\beta$$

The factor 2 vanishes and

$$d\nu = \frac{m_o c^2}{h}(1-\beta^2)^{-3/2}\beta d\beta \tag{2-15}$$

An expression for dk is obtained for use in Equation 2-13 in a similar way. Starting with Equation 2-5 rearranged in form, and again using the relativistic expression for mass, gives

$$\frac{1}{\lambda} = \frac{p}{h} = \frac{mv}{h} = \frac{m_o v}{h(1-\beta^2)^{1/2}} \tag{2-16}$$

Then, recalling that $v = c\beta$, and that $1/\lambda = k$, Equation 2-16 is reexpressed as

$$k = \frac{1}{\lambda} = \frac{m_o c}{h} \cdot \frac{\beta}{(1-\beta^2)^{1/2}} \tag{2-17}$$

Upon differentiation, Equation 2-17 becomes

$$dk = \frac{m_o c}{h}\left[\beta\left(-\frac{1}{2}\right)\cdot(1-\beta^2)^{-3/2}(-2\beta) + (1-\beta^2)^{1/2}\right]d\beta$$

The factor (−2) vanishes and

$$dk = \frac{m_o c}{h}\left[\beta^2(1-\beta^2)^{-3/2} + (1-\beta^2)^{-1/2}\right]d\beta$$

Factoring the quantity $(1 - \beta^2)^{-3/2}$ will result in an equation similar to Equation 2-15 for $d\nu$

$$dk = \frac{m_o c}{h}(1-\beta^2)^{-3/2}\left[\beta^2 + (1-\beta^2)\right]d\beta$$

This reduces to

$$dk = \frac{m_o c}{h}(1-\beta^2)^{-3/2}\,d\beta \tag{2-18}$$

Expressions now are available to investigate implications of Equation 2-13.

Equations 2-15 and 2-18 are substituted into Equation 2-13 and result in

$$v_g = \frac{d\nu}{dk} = \frac{\dfrac{m_o c^2}{h}\beta(1-\beta^2)^{-3/2}\,d\beta}{\dfrac{m_o c}{h}(1-\beta^2)^{-3/2}\,d\beta} = c\beta \tag{2-19}$$

But, $\beta = v/c$. So,

$$v_g = c\frac{v}{c} = v \tag{2-20}$$

Thus, when the group velocity equals the phase velocity the wave packet may be treated like a particle. The phase velocity has no direct role in classical mechanics. However, under the conditions noted above, this equivalence reduces waves to their classical counterparts. Such "particles" can be treated classically if they obey Equation 2-20. If they do not obey this equation, they "spread out" as they move and cannot be treated as particles. This applies to matter waves associated with a particle of matter, a quantum of energy, or with the corpuscular properties of electromagnetic radiation.

Thus, it has been established that electrons can be treated either as particles or as matter waves. The same is true for the electromagnetic radiation described above. The ability of being treated in either of these ways is known as duality. This situation arises not from the electrons or the light waves themselves, but from the inability to describe their behavior more precisely.

It is interesting to note again that this dualism is implicit in Einstein's fourth assumption (see Section 1.3) that photons behave like waves of corresponding frequency.

In general, the longer the wavelength, the more difficult it is to show the corpuscular nature of the radiation. The wave-like character is not pronounced because of the low energies of the quanta. As the wavelength becomes shorter the corpuscular properties become more evident.

2.2. THE DE BROGLIE WAVELENGTH

Use was made of the de Broglie equation in the preceding section prior to proving it. This important contribution, anticipated by Einstein, is derived in this section.

The relationship between particle- and wave-like behavior was given by de Broglie (1924). In effect, he suggested that a wavelength could be associated with a particle and that that wavelength is related to the momentum of the particle, p. This is given as

$$\lambda = \frac{h}{p} \tag{2-21}$$

This emphasizes the dual behavior of such particles. It should be noted that other particles (photons, neutrons, alpha particles, etc.) as well as electrons show this duality.

The de Broglie relationship can be derived by starting with Equation 2-9.

$$v_g = v - \lambda\frac{dv}{d\lambda} \tag{2-9}$$

An expression for $dv/d\lambda$ is obtained from differentiating Equation 2-7 with respect to λ

$$\frac{dv}{d\lambda} = \lambda\frac{d\nu}{d\lambda} + \nu \tag{2-22}$$

The substitution of Equation 2-22 into Equation 2-9 gives

$$v_g = v - \lambda\left(\lambda\frac{d\nu}{d\lambda} + \nu\right)$$

This is reexpressed as

$$v_g = v - \lambda^2 \frac{d\nu}{d\lambda} - \lambda\nu$$

Since $\lambda\nu = v$ (Equation 2-7) its substitution results in

$$v_g = v - \lambda^2 \frac{d\nu}{d\lambda} - v = -\lambda^2 \frac{d\nu}{d\lambda} \quad (2\text{-}23)$$

This may be rearranged as

$$\frac{d\nu}{d\lambda} = -\frac{v_g}{\lambda^2} \quad (2\text{-}24)$$

This will be used in Equation 2-26, below. Now, by equating the classical and quantum expressions for energy,

$$E = \frac{1}{2} mv^2 = h\nu \quad (2\text{-}25)$$

and differentiating this with respect to wavelength,

$$mv \frac{dv}{d\lambda} = h \frac{d\nu}{d\lambda}$$

or, upon rearrangement,

$$\frac{dv}{d\lambda} = \frac{h}{mv} \cdot \frac{d\nu}{d\lambda} \quad (2\text{-}26)$$

Now, by means of Equation 2-24, Equation 2-26 becomes

$$\frac{dv}{d\lambda} = \frac{h}{mv}\left[-\frac{v_g}{\lambda^2}\right] \quad (2\text{-}27)$$

Now, since the phase and group velocities must be equal (Equation 2-20) so that the wave packet can be treated as a particle, then $v = v_g$ and

$$\frac{dv}{d\lambda} = \frac{h}{m}\left[-\frac{1}{\lambda^2}\right] \quad (2\text{-}28)$$

Upon integration this becomes, when the constant of integration is neglected,

$$v = \frac{h}{m\lambda}$$

or

$$\lambda = \frac{h}{mv} = \frac{h}{p} \quad (2\text{-}29)$$

which is the de Broglie equation. The wavelength obviously is best associated with a wave, the momentum with a particle. In Equation 2-29 their relationship has been derived. It is now apparent why this is useful in the description of the wave-like behav-

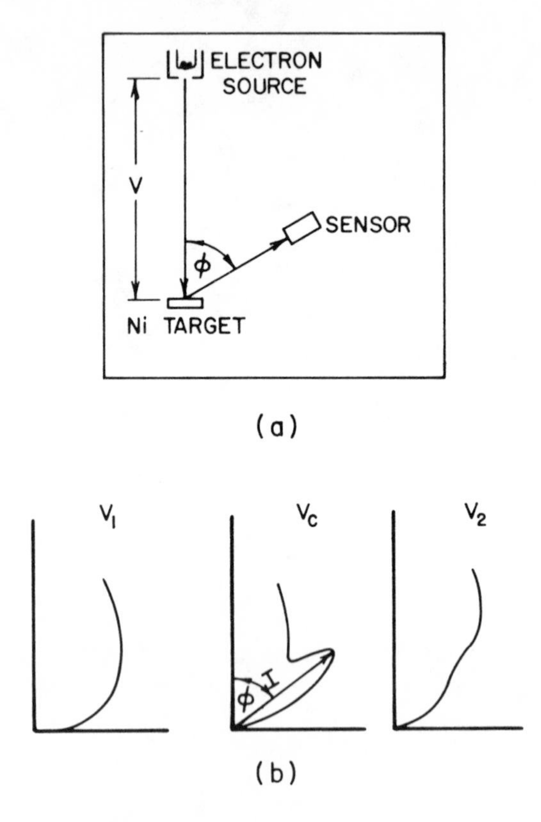

FIGURE 2-3. The Davisson-Germer experiment. (a) Schematic diagram of the vacuum chamber; (b) some polar plots of intensities; V_c = 54V.

ior which can be associated with an electron; it relates wave and particle properties in a simple way.

2.3. THE DAVISSON-GERMER EXPERIMENT

The dual nature of electrons was demonstrated by Davisson and Germer (1928) when studying the diffraction of a stream of electrons. The approach used is shown in Figure 2-3.

Electrons were accelerated from the source to the nickel target by means of a voltage between the two. The nickel target, absorbing thermal energy from the electrons impinging upon it, recrystallized into a structure of a relatively few, large grains. Each of these acted as a single crystal which served as a diffraction grating. The intensity of the diffracted beam was plotted in polar coordinates for each accelerating voltage. As this voltage was increased, a peak appeared on the curves. This reached its greatest magnitude at 54V and $\phi = 50°$; it then diminished and disappeared at higher voltage.

The increase in intensity at 50° was interpreted as constructive interference. If an electron can be treated as a wave, this should occur when Bragg's law is satisfied: $n\lambda = 2d \sin \theta$. Since this is a first-order reflection, n = 1. When the orientation of the nickel crystal is taken into account, in order to determine the interplanar distance, d, of the diffracting planes, it was found that $\lambda = 1.65$ Å.

If, instead of waves, the electron beam is considered to be a stream of particles, and if the de Broglie relationship is correct, it should be possible to compute the wavelength associated with such electrons. The kinetic energy of an electron can be expressed by

$$K.E. = \frac{1}{2} mv^2 = \frac{eV}{300}$$

(The factor 1/300 is used to convert volts to statvolts.) This can be rewritten as

$$mv^2 = \frac{2eV}{300} = \frac{eV}{150}$$

Both sides of this equation are multiplied by the mass, m,

$$m^2v^2 = \frac{meV}{150}$$

and the momentum is found to be

$$p = mv = \left[\frac{meV}{150}\right]^{1/2}$$

Now, making use of de Broglie's equation

$$\lambda = \frac{h}{p} = h\left[\frac{150}{meV}\right]^{1/2}$$

When this calculation is carried out, it is found that $\lambda = 1.67$ Å.

This high degree of agreement is considered to verify the de Broglie equation and to confirm the concept of duality.

2.4. THE COMPTON EFFECT

The reflection or diffraction of a beam of electromagnetic radiation can be expected to change its intensity, but changes in its wavelength would not ordinarily be anticipated.

Compton (1923) showed that when monochromatic X-rays are scattered by a solid, the resultant beam will show an additional different wavelength, or frequency, instead of just the original one. This is shown schematically in Figure 2-4. The peak at λ_o is that of the original beam of X-rays. The additional peak at the longer wavelength represents the result of a reaction between the X-rays and the relatively free electrons in the solid. This was explained by considering the change in the momentum of a photon of the X-rays when it impinges upon a valence electron and imparts momentum, p_e, to it. This is shown in Figure 2-5.

When conservation of energy and momentum are taken into account, the shift of wavelength can be explained. This shift depends upon the scattering angle, but is independent of the wavelength and the material doing the scattering. This phenomenon clearly shows that radiation treated as a particle can impart momentum to a particle of matter (the valence electron).

Since some of the incident radiation is unchanged by this process, another reaction must take place. Here the incident photons interact with the electrons which are much more strongly attached to ions of the solid in the form of completed shells. The great difference in the masses involved between a photon and an ion core is responsible for

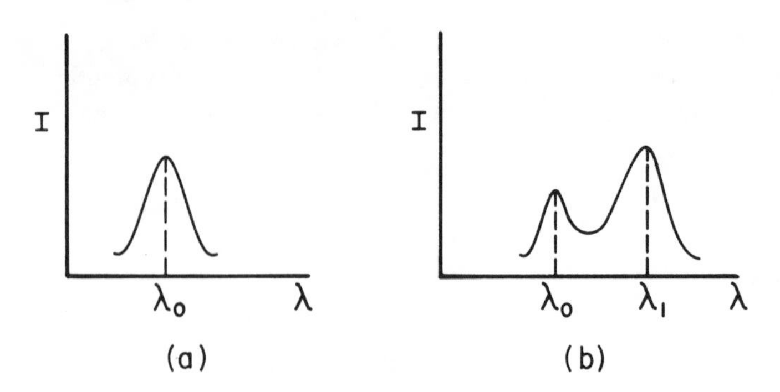

FIGURE 2-4. Intensities of X-rays (a) before and (b) after scattering. $\Delta\lambda = \lambda_1 - \lambda_o = h/mc\ (1 - \cos\theta)$. See Figure 2-5.

the virtually unchanged momentum of the scattered photon. Thus, any change in its wavelength is negligible. This accounts for the presence of the diminished peak which appears at virtually the same wavelength as the original unscattered beam.

The changes in the properties of a photon and an electron after collision are important considerations in understanding the Heisenberg Uncertainty Principle.

2.5. SPACE AND TIME LIMITATIONS

The fact that electrons and photons show dual behavior makes it possible to explore some of their properties by means of diffraction phenomena where the classical theory cannot explain the observed results, and to examine them by means of the wave theory. Here, it will become apparent that the accuracy by which some of the parameters can be known becomes increasingly smaller as the attempt is made to know others more precisely.

Consider the diffraction phenomenon caused by the passage of a beam of light, made up of an uninterrupted stream of photons, through a slit (Figure 2-6).

This diffraction behavior can only be explained by wave theory. The angle between the maximum of intensity and the first minimum is θ. The condition for this behavior is

$$\sin\theta = \frac{\lambda}{a} \simeq \theta \tag{2-30}$$

for small values of θ. This relationship of wavelength to slit size constitutes a space limitation for the following reasons: where the slit opening, a, is very much larger than the wavelength, λ, no interference, and hence essentially no diffraction will occur. A ray, or beam of light, will be virtually unaffected by the opening; its optical behavior may be predicted by classical means. But, when the slit opening is of the correct size with respect to λ, diffraction takes place and geometric optics can no longer be used. Further reduction in a, the slit opening, causes more pronounced diffraction, rather than the isolation of a ray, so that Equation 2-30 constitutes a space limitation. This also holds for electrons, since matter waves behave similarly.

What happens when the number of waves, or photons, passing through the slit is limited? Now suppose that a shutter is placed behind the slit and that a detector which is sensitive to small changes in λ, or frequency, ν, is placed so that it can scan the screen. What happens when the shutter is opened for a very short time, Δt? The short grouplet of waves, or photons, which passes through will have been changed by this. This may be shown as described below.

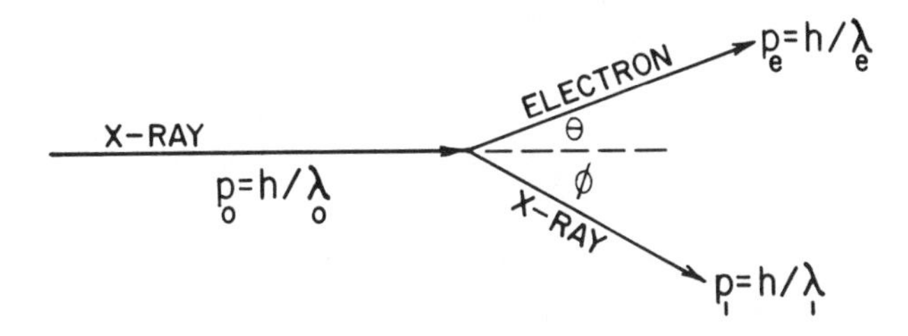

FIGURE 2-5. Schematic momenta vectors involved in the Compton effect.

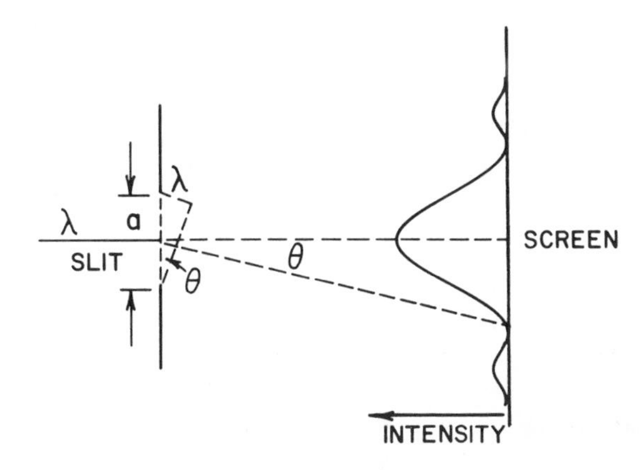

FIGURE 2-6. Diffraction of monochromatic light by a slit.

The phase velocity of the incident waves is given by Equation 2-7. Upon differentiation and rearrangement this becomes

$$\left|\frac{d\nu}{\nu}\right| = \left|\frac{d\lambda}{\lambda}\right|$$

or

$$\frac{\Delta\nu}{\nu} = \frac{\Delta\lambda}{\lambda}$$

The quantity $\Delta\nu/\nu$ is the reciprocal of the number of waves in the grouplet. Another expression for the number of waves in this very short train is given by $v\Delta t/\lambda$. Both of these expressions may be employed to give

$$\frac{\Delta\nu}{\nu} = \frac{\lambda}{v\Delta t} \qquad (2\text{-}31)$$

Equation 2-7 may be used for the phase velocity in Equation 2-31 to obtain

$$\frac{\Delta\nu}{\nu} = \frac{\lambda}{\lambda\nu\Delta t}$$

or

$$\Delta\nu = \frac{1}{\Delta t} \qquad (2\text{-}32)$$

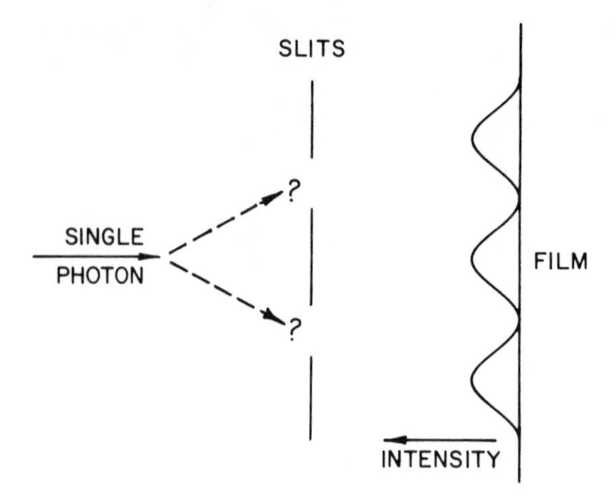

FIGURE 2-7. Schematic sketch of the G.I. Taylor experiment.

Thus, the frequency of the waves comprising the short train is no longer a constant but varies over a range of frequencies, $\Delta\nu$. This range increases as the open time of the shutter decreases. Thus, the closer the attempt to examine the short group of waves, the greater the frequency range spreads and becomes increasingly uncertain, thus introducing a time limitation.

An uninterrupted stream of photons was used to describe the space limitation. The use of a limited number of photons introduced a time limitation. What happens when just one photon at a time passes through a pair of slits? This is shown in Figure 2-7.

The intensity can be decreased so that there is a high probability of finding only one photon between the first slit and the film at any given time. If this is permitted to continue sufficiently long, so that the sum of such photons approximates the number which would reach the film in a given time at more usual intensities, the diffraction pattern is found to be the same as that caused by plane waves of light passing through *two* slits.

If a photon is a particle, it should only be able to pass through one slit, not both. The resultant diffraction pattern should then be quite different from that obtained. The observed pattern can only be caused by the passage of a photon through both slits. How could the photon have passed through both slits at the same time? It would thus appear that a photon is different from the concepts presented here earlier for either particles or waves. It turns out that it is not possible to know the exact nature of a photon when an attempt is made to examine it closely. It may also be concluded that the same is true for an electron as well.

2.6. HEISENBERG UNCERTAINTY PRINCIPLE

It has been shown that photons and electrons may be described in terms of waves and that this approach has limitations. The closer the attempt is made to study the behavior of a photon or a particle, the greater will be the error of the measurements. Heisenberg summarized this in his Uncertainty, or Indeterminacy, Principle. This concept was demonstrated by means of idealized "thought" experiments.

If it is desired to "see" an electron in such an experiment, the photon must enter the microscope so that the electron may be observed. Both the photon and the electron interact in the same way as described to explain the Compton effect (see Section 2.4).

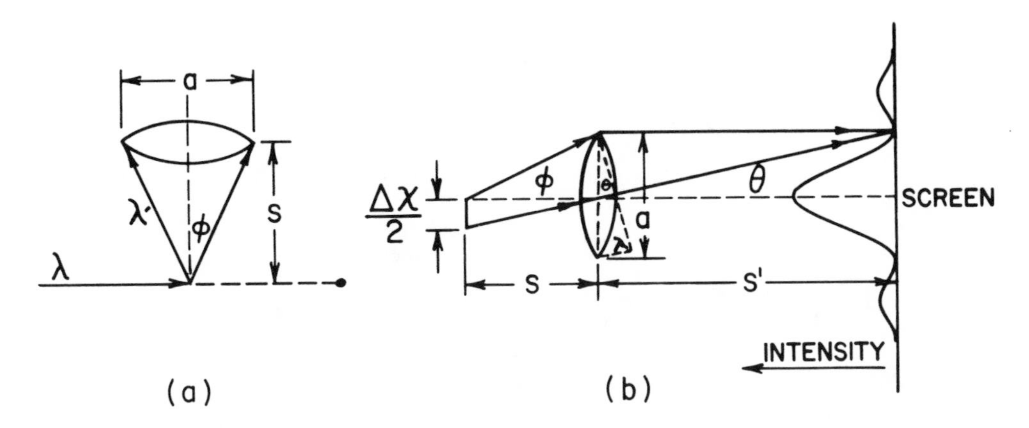

FIGURE 2-8. Diagrams for the Heisenberg "thought" experiment. (a) The photon scatters the electron and then enters the "microscope"; (b) optics involved. This combines Figures 2-6 and 2-8a.

The photon will undergo a change in momentum and will be diffracted if and when it enters the microscope. Knowledge of the position of the electron will depend upon the diffraction pattern of the photon (Figure 2-8), and the degree of fineness with which the microscope can resolve an object. It is most probable that the photon will fall within the central maximum of the diffraction pattern. However, the sharpness of the image of the electron, which is projected upon the screen, will be affected by the small maxima which appear on either side of the central maximum; the outline of the image will be fuzzy. This occurs because there are much smaller, but equally valid, probabilities that the photon will fall in any of the smaller intensity maxima which appear on both sides of the central maximum.

A photon "reflected" from the electron, having undergone a momentum change, can enter the objective lens of the microscope at any angle between $\pm\phi$ (Figure 2-9). The maximum change of its momentum in the x direction is

$$\Delta p_x = 2p \sin \phi$$

and, for small angles,

$$\Delta p_x \simeq 2p\,\phi$$

This gives

$$\phi \simeq \frac{\Delta p_x}{2p}$$

The degree with which any microscope can form a sharply defined image is limited by

$$\Delta x = \frac{\lambda}{2n \sin \phi}$$

Here Δx is the fineness with which an object may be resolved and n is the index of refraction of the medium in which the light travels in passing from the object being viewed to the objective lens of the microscope. The degree of resolution is indicated as being centered on the optical axis, but, for simplicity, only half of this, $\Delta x/2$, is shown in Figure 2-8b. The medium in which the photon travels is a vacuum, so n =

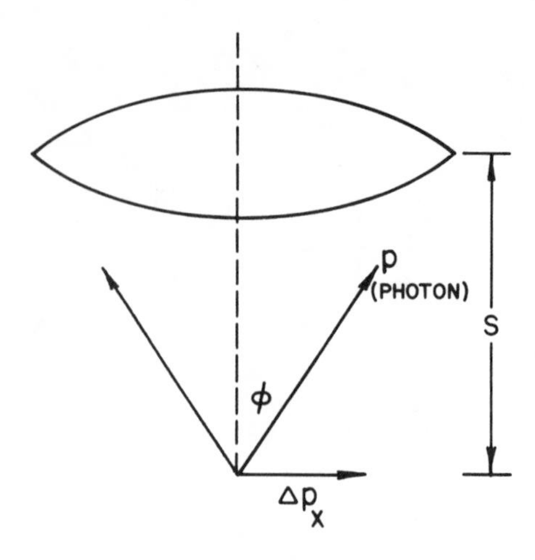

FIGURE 2-9. Momentum diagram for the Heisenberg "thought" experiment, considering only the photon.

1. Thus, where ϕ is small,

$$\Delta x = \frac{\lambda'}{2\phi}$$

So that

$$\phi = \frac{\lambda'}{2\Delta x}$$

The two expressions for ϕ are equated, giving

$$\frac{\Delta p_x}{2p} = \frac{\lambda'}{2\Delta x}$$

The coefficient 1/2 vanishes and this equation may be rewritten as

$$\Delta p_x \, \Delta x = \lambda' p$$

The de Broglie relationship, Equation 2-29, is used to reexpress λ'. This gives

$$\Delta p_x \, \Delta x = \frac{h}{p} p = h$$

More precisely, this equation should be written as

$$\Delta p_x \, \Delta x \geq h \tag{2-33}$$

Figure 2-8 also may be used to obtain Equation 2-33 by another means. This approach involves the use of tan θ and tan ϕ in conjunction with a suitable expression for Δp_x, when the space limitation, Equation 2-30, is taken into consideration (Problem 6).

Any situation which changes the momentum of a photon would be expected to affect its energy. The energy uncertainty can be shown in an elementary way starting with Equation 2-3

$$E = h\nu$$

and differentiating it to obtain

$$\Delta E = h\Delta\nu \quad (2\text{-}34)$$

The time limitation, Equation 2-32, is used in Equation 2-34 to find

$$\Delta E = h \frac{1}{\Delta t}$$

or, more precisely,

$$\Delta E \Delta t \geq h \quad (2\text{-}35)$$

Thus, the Heisenberg Uncertainty Principle places limits upon the degree to which the products of the conjugate parameters in Equations 2-33 and 2-35 can be known. The accuracies of the measurements of x and p or of E and t are not limited in themselves. However, no pair of these conjugates can be equally well-known simultaneously. The greater the accuracy with which one is known, the lesser must be the accuracy with which the other is known.

The classical mechanics assumes that position, momentum, energy, and time can be known exactly. It contains no analogue of Heisenberg's principle. This is not the case in quantum mechanics; these parameters cannot be known exactly. Indeed, if one of them is known exactly, complete uncertainty must exist with regard to its conjugate. Since the precision allowed in classical mechanics is not permitted in quantum mechanics, the best that can be done is to make use of the most likely values. It is for this reason that the quantum mechanics must be based upon a statistical approach which provides the most probable values.

2.7. STATISTICAL APPROACH

The properties of the physical entities which cannot be described by the classical mechanics include the following basic considerations: electromagnetic radiation may be treated as photons; photon energy is exchanged in discrete amounts; photons can have momentum; matter particles can show dual behavior and may be treated in terms of wave properties; and, limits affect the degree of precision with which these things can be known by classical means. The quantum mechanics, therefore, must take into account this particular set of characteristics. As has been shown in the preceding sections, this can be done in ways which describe their most probable characteristics.

A diffraction pattern will be used to provide an insight into the statistical approach to the quantum mechanics. While an incident beam of photons is used in the following explanation, it must be remembered that identical results may be obtained for any quantized, indistinguishable particle, such as an electron, a proton, a neutron, etc., which shows dual behavior (see Section 5.4 in Chapter 5).

A circular aperture with the same diameter, a, as the small slit shown in Figure 2-6 is used here for simplicity. A beam of monochromatic light passing through this slit will result in a circular diffraction pattern. This diffraction pattern and its intensity as a function of θ are shown in Figure 2-10a and b as taken along any diameter of the pattern.

Upon passing through the aperture, a photon may be diffracted in any direction in space. This becomes increasingly apparent as the ratio of a/λ approaches zero (see

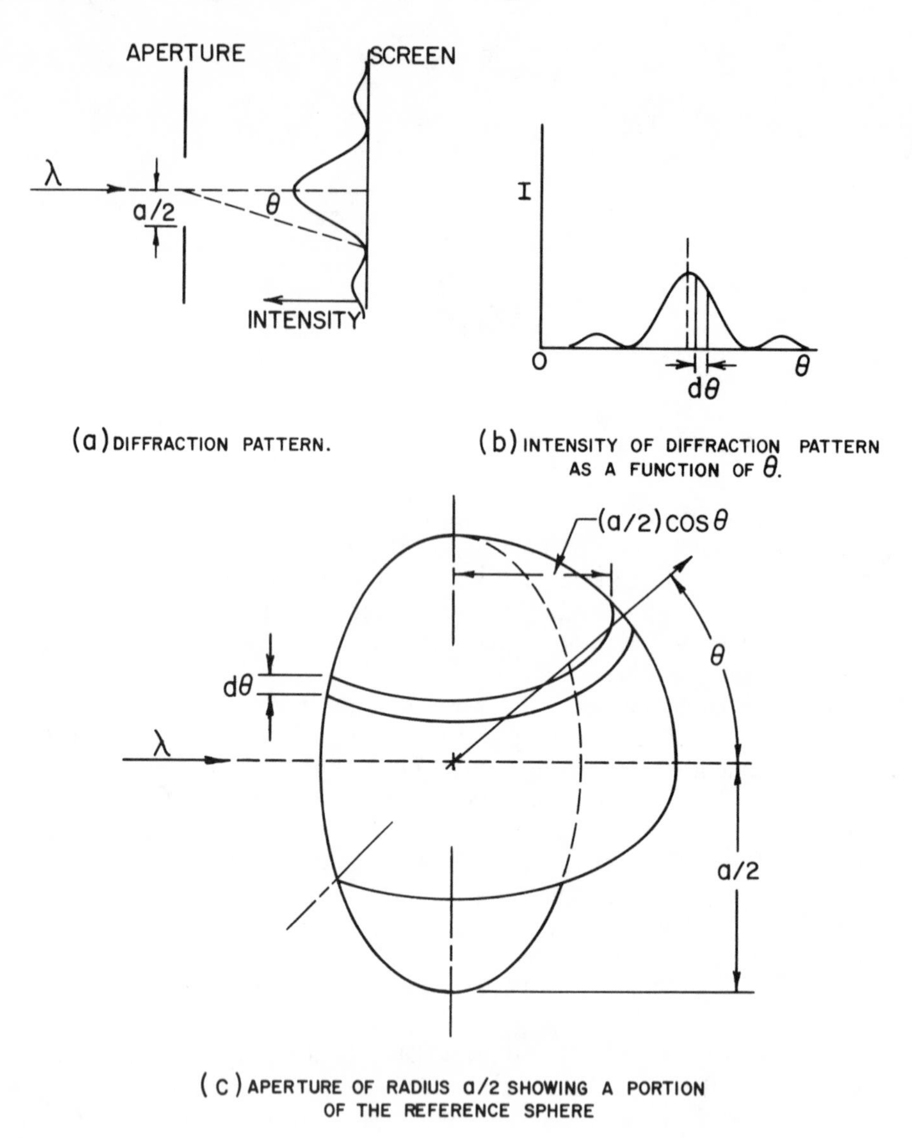

FIGURE 2-10. Diffraction pattern of a circular aperture for approximating the photon flux.

Section 2.5). An approximation of the number of photons being diffracted per unit area per unit time, or the flux, f, is made by reference to the hemisphere indicated in Figure 2-10c. The radius of the hemisphere is a/2 so that its area is $\pi a^2/2$. The element of hemispheric surface area is $\pi r d\theta = \pi(a/2) \cos\theta \, d\theta$. Let N be the number of photons which cross the surface of the hemisphere in unit time. Then, dN will be the number of photons which cross the element of surface area in unit time. The ratio of dN/N will be given by the ratio of their respective areas:

$$\frac{dN}{N} = \frac{\pi(a/2)\cos\theta\, d\theta}{\frac{\pi}{2}a^2} = \frac{\cos\theta\, d\theta}{a}$$

since $\pi a/2$ vanishes and

$$dN = \frac{N \cos \theta \, d\theta}{a}$$

The flux across a plane perpendicular to the optical axis and parallel to the planes of the aperture and the screen is given by

$$df = dN \sin \theta$$

or, upon substitution for dN,

$$df = \frac{N \sin \theta \cos \theta \, d\theta}{a}$$

The intensity of the pattern is greatest in the region of the central maximum. So, for the small angle, it may be approximated that $\sin \theta \simeq \theta$ and $\cos \theta \simeq 1$, giving

$$df \simeq \frac{N \theta \, d\theta}{a}$$

Use now is made of the space limitation, Equation 2-30, to substitute for θ:

$$df \simeq \frac{N}{a} \cdot \frac{\lambda}{a} d\theta = \frac{N \lambda \, d\theta}{a^2}$$

The implicit coefficient unity should be represented by an integer α which determines the minima in the diffraction pattern. Thus, more generally,

$$df \simeq \frac{\alpha \lambda N \, d\theta}{a^2}$$

The fraction N/a^2 is the number of photons per unit time per unit area or the intensity, I, of the photons. Therefore,

$$df \simeq \alpha \lambda I \, d\theta$$

and the probability of finding a photon anywhere on the screen will vary as df. Since prime interest is attached to the greatest probability of finding a photon from the monochromatic beam, and the intensity is highest in the central maximum, then the optimum likelihood of finding a photon will vary as

$$df \simeq \text{const } I \, d\theta$$

A means is now sought to translate the above findings into a more general and useful form. This may be done by means of a function, ψ, which may be regarded as giving the *amplitude* of a de Broglie wave, and which is called a wave function. This wave function is meaningless by itself, especially since most wave functions are complex; that is, they involve $(-1)^{1/2}$ which is denoted by i. It was postulated by Born that the *intensity* of a wave could be given by $\psi^*\psi$ where ψ^* is the complex conjugate of ψ. Thus, if $\psi = a + ib$, $\psi^* = a - ib$, then $\psi^*\psi = a^2 + b^2$; the product of the conjugate amplitudes gives a real intensity which does have meaning.

This concept may be applied to the probability of finding a photon. On the basis of the foregoing, the intensity of the diffraction pattern may be expressed as

$$I \propto \psi^*\psi$$

and

$$I d\theta \propto \psi^*\psi \, d\theta$$

Or,

$$df = \text{const } I d\theta = C^2 \psi^*\psi \, d\theta$$

This gives the probability of finding a photon in the element of $d\theta$ in terms of the amplitude of a de Broglie wave. This is frequently written as

$$df = \text{const } I d\theta = C^2 |\psi|^2 \, d\theta$$

In both of the above expressions C is a constant.

These equations were derived on the basis of the most probable range in which a photon would be found. But this represents only a fraction of the entire range in which it is possible to find a photon. The probability that a photon will be found anywhere on the screen is

$$B^2 \int_{-\pi/2}^{\pi/2} \psi^*\psi \, d\theta$$

where B^2 is an arbitrary constant selected such that

$$B^2 \int_{-\pi/2}^{\pi/2} \psi^*\psi \, d\theta = 1$$

This is called normalization. The probability of finding a photon in the most probable range with respect to the entire range is

$$\frac{C^2 \psi^*\psi \, d\theta}{B^2 \int_{-\pi/2}^{\pi/2} \psi^*\psi \, d\theta} = C^2 \psi^*\psi \, d\theta$$

This is called the probability density. It may be applied to any range of interest.

Since Heisenberg's Uncertainty Principle states that the position of the photon cannot be known precisely, the only alternative is to find its most likely location. In the above attempt to do this, it must be emphasized that this is a probability and that the photon may be anywhere on the screen, in positions ranging from $\pi/2$ to $-\pi/2$. The fact that the intensity is greatest in the central maximum was used above to simplify the approximation. This now takes on greater significance because it means that the probability densities of properly selected wave functions will be greatest in that region.

The ideas about wave functions outlined here will be used in the development of Schrödinger's equation, Chapter 3, and elsewhere.

2.8. PROBLEMS

1. Show by graphical means how two waves of slightly different wavelengths can interfere to form a wave packet.
2. Calculate the momenta of particles with the following wavelengths: 1 Å, 10 Å, 100 Å, 1000 Å.
3. Verify the Davisson-Germer findings for a diffracting crystal of nickel whose orientation perpendicular to the beam is (111).
4. Calculate the kinetic energies and momenta of electrons accelerated by 10^2, 10^4, and 10^6 V.
5. What slit opening should be used to ensure that diffraction would take place if monochromatic light of $\lambda = 0.6869\mu$ was to be used?
6. Derive Equation 2-33 by the method indicated in Section 2.6.
7. Make a graph of one of the Heisenberg relationships. What class of curve is this? What are its implications?

2.9. REFERENCES

1. **Sproull, R. L.,** *Modern Physics,* John Wiley & Sons, New York, 1956.
2. **Richtmyer, F. K., Kennard, E. H., and Lauritsen, T.,** *Introduction to Modern Physics,* 5th ed., McGraw-Hill, New York, 1955.
3. **Hume-Rothery, W.,** *Atomic Theory for Students of Metallurgy,* The Institute of Metals, London, 1952.
4. **Dekker, A. J.,** *Solid State Physics,* Prentice-Hall, Englewood Cliffs, N.J., 1959.
5. **Kittel, C.,** *Introduction to Solid State Physics,* 3rd ed., John Wiley & Sons, New York, 1966.
6. **Wood, R. M.,** *Physical Optics,* Dover, New York, 1967.
7. **Sears, F. W.,** *Principles of Physics III, Optics,* Addison-Wesley, Reading, Mass., 1945.

Chapter 3

THE SCHRÖDINGER WAVE EQUATION

The approach here will be to start with some elementary ideas of classical mechanics and to convert these concepts to their equivalents in quantum mechanics. This will be done by changing the total energy of an electron in classical terms into its quantum mechanical equivalents and arriving at the Schrödinger equation. The solutions to this equation will provide expressions for ψ which will result in naturally discrete energy values for the particles being considered.

The solutions to the Schrödinger equation obtained here will be applied to describe the behaviors of electrons and ions. More complex applications of Schrödinger's equation will be outlined for the hydrogen and helium atoms. These will show the basis for the Pauli exclusion principle and explain why the quantum numbers which are the same as those obtained for the hydrogen atom may be employed for atoms with many more electrons. The solutions for electrons will be used to describe the electron configurations of atoms, and in subsequent chapters to develop the fundamental ideas of the modern theories of solids. Those for ions will be used primarily to describe their contributions to the thermal properties of solids.

3.1. CLASSICAL MECHANICS

A brief review of some of the classical ideas is presented as an introduction to the quantum mechanics. Consider a mechanical system in which a particle of mass, m, is acted upon by a force. For simplicity, the particle will be considered to move in the x direction. The force, then, is given by F(x). This force function is also defined as

$$F(x) = -\frac{d}{dx} V(x) \tag{3-1}$$

where V(x) is a potential function.

The following Newtonian relationships are given, on this basis, as a review. The familiar relationship between force, mass, m, and acceleration, a, is given by

$$F(x) = m\frac{d^2x}{dt^2} = ma \tag{3-2}$$

The momentum, p, is given in terms of velocity, v, by

$$p = m\frac{dx}{dt} = mv \tag{3-3a}$$

or

$$\frac{dx}{dt} = \frac{p}{m} \tag{3-3b}$$

Differentiation of Equation 3-3a with respect to time gives

$$\frac{dp}{dt} = m\frac{d^2x}{dt^2} = m\frac{dv}{dt} = ma = F(x) = -\frac{dV(x)}{dx} \tag{3-4}$$

The kinetic energy, K.E., of the particle is given by

$$\text{K.E.} = \frac{1}{2}mv^2 = \frac{p^2}{2m} \tag{3-5}$$

It should be noted that this is a function of motion, whereas the potential energy, P.E., is one of position:

$$\text{P.E.} = V(x) \tag{3-6}$$

The total energy of the system, E, is

$$E = \text{K.E.} + \text{P.E.} = \frac{1}{2}mv^2 + V(x) = \frac{p^2}{2m} + V(x) \tag{3-7}$$

A function may be defined as the total energy of the system in the terms of the variables x and p. This is called the Hamiltonian function of the system and is written as

$$H(x,p) = H = \frac{p^2}{2m} + V(x) \equiv E \tag{3-8}$$

where E is the total energy of the system. For simplicity H(x,p) will be written as H. From Equation 3-8, differentiating with respect to momentum,

$$\frac{\partial H}{\partial p} = \frac{p}{m} \tag{3-9}$$

Since both Equations 3-3b and 3-9 equal p/m,

$$\frac{p}{m} = \frac{dx}{dt} = \frac{\partial H}{\partial p} \tag{3-10}$$

The differentiation of the Hamiltonian function, Equation 3-8, with respect to x is

$$\frac{\partial H}{\partial x} = \frac{dV(x)}{dx} \tag{3-11}$$

so that by Equation 3-4

$$-\frac{dV(x)}{dx} = \frac{dp}{dt} = -\frac{\partial H}{\partial x} \tag{3-12}$$

Equations 3-10 and 3-12 constitute the Hamiltonian equivalents of the classical expressions given by Equations 3-3 and 3-4. In this way, the momentum and force functions are replaced by the appropriate forms of the Hamiltonian functions which define the variables x and p.

The Hamiltonian function for the total energy of a system will be used as a basis for obtaining expressions from which Schrödinger's equation can be obtained. The solutions to this equation will form the basis for understanding the topics which follow.

3.2. APPLICATION TO QUANTUM MECHANICS

The variables, position, and momentum, in the Hamiltonian functions can be determined sufficiently well for classical mechanical applications. This is not the situation for the quantum mechanic case. The Heisenberg Uncertainty Principle (Section 2.5 in Chapter 2) places limits upon the certainty to which the position-momentum and energy-time conjugates may be known. Therefore, as previously indicated, the approach is based upon the postulates that the most probable values can be determined by the wave function and that the Hamiltonian for the system can be expressed as in Equation 3-8. This equation is transformed into the wave equation by the replacement of the variables by differential operators.

Let the momentum of the particle be represented by the operator such that it becomes, where $i = \sqrt{-1}$,

$$P \rightarrow \frac{h}{2\pi i}\frac{\partial}{\partial x} \tag{3-13}$$

and the total energy becomes

$$E \rightarrow -\frac{h}{2\pi i}\frac{\partial}{\partial t} \tag{3-14}$$

It will be observed that each of these relationships involves conjugates. Its position, x, is not a function of a conjugate and remains untransformed:

$$x \rightarrow x \tag{3-15}$$

Now consider the Hamiltonian function for one dimension

$$H \equiv H(x,p) = \frac{p^2}{2m} + V(x) \equiv E \tag{3-8}$$

It will be recalled from Section 2.6 in Chapter 2 that, in dealing with these variables, an approach based upon probability must be employed. The wave function, Ψ, (previously defined as a function of space and time) was employed in such a way that $\Psi^*\Psi$ was proportional to the probability of finding the particle. This can be applied more generally. The function ψ may be made to represent any observable quantity. Let Ψ be a function of position, x, and time, t. In the present case momentum, position, time, and energy are the factors involved. The quantum mechanic representation is obtained for each of these variables by performing the suitable operations on ψ given by Equations 3-13, 3-14, and 3-15 starting with Equation 3-8 as follows:

$$H(x,p) = E \tag{3-8}$$

and allowing H(p,x) and E to operate on $\Psi(x,t)$ changes Equation 3-8 to

$$H(x,p)\Psi(x,t) = E\Psi(x,t) \text{ or } H\Psi = E\Psi \tag{3-8a}$$

The wave function often is expressed simply as Ψ, for convenience.

The operation to determine $p^2/2m$ requires that the operator be used twice because the momentum is squared. This is done as follows:

$$\frac{1}{2m}\left[\frac{h}{2\pi i}\frac{\partial}{\partial x}\right]\left[\frac{h}{2\pi i}\frac{\partial}{\partial x}\right]\Psi = -\frac{1}{2m}\frac{h^2}{4\pi^2}\frac{\partial^2\Psi}{\partial x^2} \tag{3-16}$$

The operation to determine V(x) is

$$V(x)\,\Psi \tag{3-17}$$

since x is unchanged. Similarly, E is determined by

$$E = \frac{-h}{2\pi i}\,\frac{\partial \Psi}{\partial t} \tag{3-18}$$

These operations transform Equations 3-8 into

$$-\frac{h^2}{8\pi^2 m}\,\frac{\partial^2 \Psi}{\partial x^2} + V(x)\Psi = -\frac{h}{2\pi i}\,\frac{\partial \Psi}{\partial t}$$

or

$$\frac{h^2}{8\pi^2 m}\,\frac{\partial^2 \Psi}{\partial x^2} - V(x)\Psi = \frac{h}{2\pi i}\,\frac{\partial \Psi}{\partial t} \tag{3-19}$$

Since this is only for the dimension, x, Ψ is a function of x and t only, rather than of x, y, z, and t. Equation 3-19 is a one-dimensional form of the Schrödinger equation for a single particle. This is also written as

$$-\frac{h^2}{8\pi^2 m}\,\frac{\partial^2 \Psi}{\partial x^2} + V(x)\Psi = E\Psi \tag{3-20}$$

The properties of the factors of this equation are of interest: the mass relates to a particle; Ψ, being an amplitude, relates to a wave; the potential function represents an external influence; and the energy, according to the Heisenberg Uncertainty Principle, cannot be known exactly.

Equation 3-20 also may be written as

$$H\Psi = E\Psi = -\frac{h}{2\pi i}\,\frac{\partial \Psi}{\partial t}$$

or

$$H(x,p)\Psi = E\Psi = -\frac{h}{2\pi i}\,\frac{\partial \Psi}{\partial t} \tag{3-21}$$

The problem becomes one of finding suitable expressions for Ψ which constitute solutions for Schrödinger's equation. This can be done only when certain criteria are satisfied (Section 3.3).

However, Equation 3-20 will serve to provide a simple illustration of the use of the Schrödinger equation. Consider the case of a free electron. All of its energy can be considered to be kinetic, hence the potential V(x) will be zero. The equation now becomes

$$-\frac{h^2}{8\pi^2 m}\cdot\frac{\partial^2 \Psi}{\partial x^2} = E\Psi \tag{3-22}$$

When the total energy of the electron is constant, a solution to this equation can be obtained by setting

$$\Psi = e^{i\bar{k}\cdot x} \tag{3-23}$$

in which $\bar{k}$ is the wave vector of the electron, and is equal to $2\pi/\lambda$, where x denotes its position. Upon differentiation of Ψ, the following are obtained:

$$\frac{\partial \Psi}{\partial x} = i\bar{k}e^{i\bar{k}\cdot x}$$

and

$$\frac{\partial^2 \Psi}{\partial x^2} = -\bar{k}^2 e^{i\bar{k}\cdot x} = -\bar{k}^2 \Psi \tag{3-24}$$

When Equation 3-24 is substituted into Equation 3-22,

$$-\frac{h^2}{8\pi^2 m}(-\bar{k}^2 \Psi) = E\Psi$$

It is seen that Ψ, Equation 3-23, is a solution to the Schrödinger equation when

$$E = \frac{h^2 \bar{k}^2}{8\pi^2 m} \tag{3-25}$$

This elementary relationship has important applications. It will be used to describe nearly "free" electron behavior in metals and will serve as a basis for the development of band structure and its implications (see Chapter 5).

This solution can be verified in a simple way. Starting with the deBroglie equation,

$$p = mv = \frac{h}{\lambda}$$

it is found that

$$v^2 = \frac{h^2}{m^2\lambda^2}$$

This may be substituted into the expression for kinetic energy to get

$$E = \frac{1}{2}mv^2 = \frac{h^2}{2m\lambda^2}$$

Use is now made of the expression for the wave vector to substitute for λ^2. When this is done

$$E = \frac{h^2}{2m}\cdot\frac{\bar{k}^2}{4\pi^2} = \frac{h^2\bar{k}^2}{8\pi^2 m} \tag{3-26}$$

Thus, Equation 3-26 verifies Equation 3-25, since identical results for the energy of a free electron were obtained by two widely differing means. Verifications such as the one given here are not normally this easy; the case presented here is the simplest possible.

The general form of Schrödinger's equation is given by

$$-\frac{h^2}{8\pi^2 m}\left[\frac{\partial^2\Psi}{\partial x^2}+\frac{\partial^2\Psi}{\partial y^2}+\frac{\partial^2\Psi}{\partial z^2}\right] + V(x,y,z)\Psi = E(x,y,z,t)\Psi \tag{3-27}$$

The amplitude Ψ is now a function of x, y, z, and t instead of just x and t, as given for the one-dimensional case.

Many applications of the Schrödinger equation deal only with the potential energy. As previously noted, this factor depends upon position; it is not a function of time. Thus, if the time factor is separated out of this equation it becomes simpler to use. The one-dimensional expressions, Equations 3-20 and 3-21, will be used to show this.

This problem is approached by using a wave function to separate the variables, which takes the form

$$\Psi(x,t) = \Psi_1(x)\varphi(t) \tag{3-28}$$

Equation 3-21 now becomes

$$\varphi(t)H(x,p)\Psi_1(x) = -\frac{h}{2\pi i}\Psi_1(x)\frac{\partial\varphi(t)}{\partial t}$$

Both sides of this equation are divided by $\Psi_1(x)$ to obtain

$$\varphi(t)\frac{H(x,p)\Psi_1(x)}{\Psi_1(x)} = -\frac{h}{2\pi i}\frac{\partial\varphi(t)}{\partial t}$$

The equation is the next divided by $\phi(t)$ and rearranged for clarity as

$$\frac{H(x,p)\Psi_1(x)}{\Psi_1(x)} = -\frac{h}{2\pi i}\cdot\frac{1}{\partial t}\cdot\frac{\partial\varphi(t)}{\varphi(t)} \tag{3-29}$$

The left-hand side of Equation 3-29 will be only a function of x after the indicated operation is performed; the right-hand side is only a function of t. Since both sides of this equation are functions of different variables and are equal to each other, they must equal a certain class of constants, E, real numbers, known as "eigenvalues". This parameter will be discussed in Section 3.5.

Attention may now be centered on the time-dependent portion of Equation 3-29. This is equated to an eigenvalue to give

$$-\frac{h}{2\pi i}\cdot\frac{1}{\partial t}\cdot\frac{\partial\varphi(t)}{\varphi(t)} = E$$

This may be rearranged as

$$\frac{\partial\varphi(t)}{\varphi(t)} = -\frac{2\pi i}{h}Edt \tag{3-30}$$

After integration this becomes

$$\phi(t) = \exp\left(-\frac{2\pi i}{h}Et\right) \tag{3-31}$$

This is the time-dependent form of Schrödinger's equation. Here the constant of integration, $\ln [\phi(t_o)]$, is arbitrarily equated to zero. This will not affect its use since $\phi(t)$ is not used alone, but with $\Psi_1(x)$ in the form $\Psi(x,t) = \Psi_1(x)\phi(t)$; and, where energy is conserved, it is not necessary to obtain an exact time dependence.

The time-independent form of the Schrödinger equation is obtained from the left-hand portion of Equation 3-29. For the reasons given above, it too is equated to the same eigenvalue used for the determination of the time-dependent equation. Thus,

$$\frac{H(x,p)\Psi_1(x)}{\Psi_1(x)} = E$$

Upon rearrangement this becomes

$$H(x,p)\Psi_1(x) = E\Psi_1(x) \qquad (3\text{-}32)$$

This is of the same form as Equation 3-8a, but with the time factor eliminated. When the original Hamiltonian function is used (Equation 3-8) the operators given by Equations 3-16 and 3-17 transform Equation 3-32 into

$$-\frac{h^2}{8\pi^2 m}\frac{\partial^2\Psi_1(x)}{\partial x^2} + V(x)\Psi_1(x) = E\Psi_1(x) \qquad (3\text{-}33)$$

This is of exactly the same form as Equation 3-20, only now it contains just one variable. This may be rewritten as

$$\frac{\partial^2\Psi_1(x)}{\partial x^2} - \frac{8\pi^2 m}{h^2}V(x)\Psi_1(x) = -\frac{8\pi^2 m}{h^2}E\Psi_1(x)$$

or

$$\frac{\partial^2\Psi_1(x)}{\partial x^2} + \frac{8\pi^2 m}{h^2}[E - V(x)]\Psi_1(x) = 0 \qquad (3\text{-}34)$$

These are time-independent forms of Schrödinger's equation.

The separation of Equations 3-20 and 3-21 into two equations, where each is a function of a single variable, Equations 3-31 and 3-34, has reduced the difficulty of their application. The principle area of interest here will be the effect of a potential field upon an electron as a function of its position in the potential field. Thus, Equation 3-34 satisfies this. It is the equation which will be of most use in this work.

3.3. CRITERIA FOR SOLUTIONS OF SCHRÖDINGER'S EQUATION

It will be recalled (Section 2.6 in Chapter 2) that Ψ was defined as a wave function whose independent variables are space and time. The probability of finding the particle in a given range, in the illustration, was shown to be proportional to $\Psi^*\Psi\, d\theta$. If the variables are changed, so that the probable location of the photon on the two-dimensional screen is of interest, rather than its angular range, the probability of finding the photon will be proportional to $\Psi^*\Psi\, dxdy$. Or, for one dimension, the probability of finding the particle will vary as $\Psi^*\Psi dx$. This one-dimensional case will be used in the following discussion.

No restrictions were expressly stated for the wave function in the illustration noted previously for the case of the free electron. However, not all of the solutions to Schrödinger's equation correspond to natural phenomena. It is therefore necessary to impose criteria upon Ψ and its derivatives in order to ensure meaningful solutions.

First, the conditions must be such that the particle is certain to be at some point within the system. To make sure that this is so, the condition

$$\int_{-\infty}^{\infty} \Psi^{*}\Psi \, dx = 1$$

must be satisfied.

Second, both Ψ and $d\Psi/dx$ must be continuous. If this were not the case, a discontinuity could exist. This could be the equivalent to the creation or destruction of particles. In the case of an electron such behavior would violate the laws of conservation.

Also, consider what would happen if Ψ provided more than one solution. This would lead to more than one probability of finding the particle under a given set of conditions. So, a third condition must be imposed that Ψ be finite and single-valued. In addition to these constraints, Ψ must never be identical to zero. That is, there must always be a real probability of finding the particle, however small that probability may be.

These constraints upon the wave functions will not only ensure that meaningful solutions to Schrödinger's equation will be obtained, but also will be of assistance in simplifying the solutions to this equation.

3.4. ONE-DIMENSIONAL WELLS

The positive charge on an ion in a metallic lattice exerts an omnidirectional electrostatic attraction for a valence electron. In effect, the positive ion acts as a three-dimensional potential "well" for the electron. The actual description of this case will be outlined in a later section. As a start, approximate one-dimensional models will be used to illustrate the way in which the Schrödinger equation is used and to begin to describe electron behavior. These will provide a basis for approximating the influence of a three-dimensional potential well upon an electron (Section 3.8).

The significant differences between the findings of the quantum and classical mechanics then will be apparent. In addition, a basis for understanding the properties of the general classes of solids will have been established.

3.5. ELECTRON IN AN INFINITE POTENTIAL WELL

The simplest case is given by the solution of the Schrödinger equation for an electron in a one-dimensional well whose sides effectively constitute infinitely high potential barriers. This is shown schematically in Figure 3-1.

According to classical physics a particle (electron) in such a well, with constant energy $E < V_o$, should move with a constant velocity (since $E = \frac{1}{2} mv^2$) and never leave the well. The probability of finding such a particle everywhere within the well is the same. The probability of finding this particle outside of the well is zero. It will be seen that these statements are quite different from some of the findings of quantum mechanics.

The well itself has been designated as region II; portions external to the well have been designated as regions I and III. The factors operative in each of the regions are given as:

- Region I: $x < 0$; $E < V$; $V = V_o$
- Region II: $0 < x < L$; $E > V$; $V = O$
- Region III: $x > L$; $E < V$; $V = V_o$

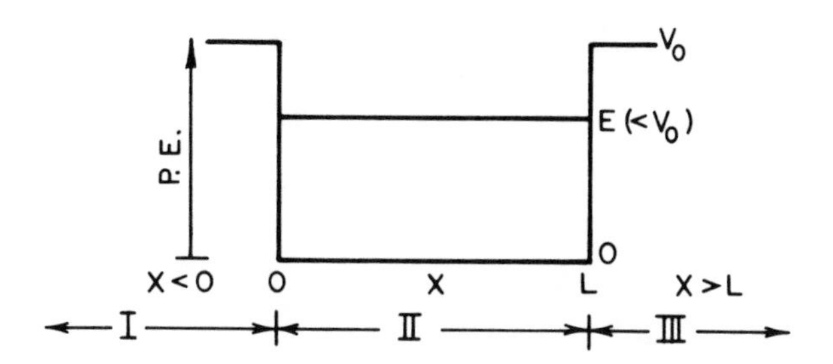

FIGURE 3-1. One-dimensional well.

These factors will be used to obtain a solution of Schrödinger's equation. Here, the influence of the potential well on the behavior of the electron is of importance. Since this is the case, the time-independent form of the Schrödinger equation, Equation 3-34, will be used. This can be expressed for this particular set of conditions as

$$\frac{\partial^2 \Psi}{\partial x^2} + \frac{4\pi^2}{h^2}\,[2m(E - V)]\,\Psi = 0 \qquad (3\text{-}35)$$

Solutions to Equation 3-35 are sought by expressing it in its general form as

$$\psi'' + R\psi = 0$$

where $R = 4\pi^2/h^2[2m(E-V)]$. Let $\psi = e^{ax}$. Then, $\psi' = ae^{ax}$ and $\psi'' = a^2e^{ax}$. The substitution of these quantities into the general equation results in

$$a^2 e^{ax} + Re^{ax} = 0$$

The exponential factors vanish and, for the case where $V > E$, R is negative:

$$a^2 - R = 0$$

A pair of equations may be based upon this equation by factoring:

$$a^2 - R = (a + R^{1/2})\,(a - R^{1/2}) = 0$$

so that

$$a = \pm R^{1/2}$$

Another pair of equations is based upon the case where $E > V$, and R is positive. This results in

$$a^2 = -R$$

so that

$$a = \pm iR^{1/2}$$

The first case, $V > E$, gives the functions

$$\psi_1 = c_1 \exp(R^{1/2}x)$$

and

$$\psi_2 = c_2 \exp(-R^{1/2}x)$$

where c_1 and c_2 are arbitrary constants. The sum of these two equations results in a solution to the general equation:

$$\psi = c_1 \exp(R^{1/2}x) + c_2 \exp(-R^{1/2}x)$$

It will be recognized that $R^{1/2} = 2\pi/h[2m(E-V)]^{1/2}$, obtained from Equation 3-35, is used in the exponential Equations 3-36 and 3-38.

The second case, $E > V$, gives another solution to the general equation:

$$\psi = c_3 \exp(iR^{1/2}x) + c_4 \exp(-iR^{1/2}x)$$

where c_3 and c_4 are arbitrary constants. It will be recognized that

$$\exp(iR^{1/2}x) = \cos(R^{1/2}x) + i\sin(R^{1/2}x)$$

and that

$$\exp(-iR^{1/2}x) = \cos(R^{1/2}x) - i\sin(R^{1/2}x)$$

These periodic, trigonometric expressions are used to give the wave function:

$$\psi = c_3 [\cos(R^{1/2}x) + i\sin(R^{1/2}x)] + c_4 [\cos(R^{1/2}x) - i\sin(R^{1/2}x)]$$

The sine and cosine terms are collected and

$$\psi = i(c_3 - c_4)\sin(R^{1/2}x) + (c_3 + c_4)\cos(R^{1/2}x)$$

Attention is now directed at the arbitrary constants. Since these are arbitrary, their difference, their sum and the product, $i(c_3 - c_4)$, must be arbitrary too. Therefore, this wave function may be written in terms of other arbitrary constants, C and D, as

$$\psi = C\sin(R^{1/2}x) + D\cos(R^{1/2}x)$$

This periodic form of the wave equation is used in Equation 3-37.

When these wave functions are used in solving Equation 3-35, $R^{1/2} = 2\pi/h[2m(E-V)]^{1/2}$ in Equations 3-36 and 3-38, where $V > E$. However, since $E > V$ and $V = O$, $R^{1/2} = 2\pi/h(2mE)^{1/2}$ in Equation 3-37. It will be recalled that E represents discrete energy levels, or eigenvalues, and V is the external potential acting upon the electron. Solutions to Equation 3-35 can now be obtained by setting ψ equal to the following appropriate wave functions:

In region I, $V > E$, $x < 0$,

$$\Psi_I = A\exp\left\{\frac{2\pi}{h}[2m(v-E)]^{1/2}x\right\}$$

$$+ B\exp\left\{-\frac{2\pi}{h}[2m(v-E)]^{1/2}x\right\} \quad (3\text{-}36)$$

In region II, V = 0, 0 < x < L,

$$\Psi_{II} = C\sin\left[\frac{2\pi}{h}(2mE)^{1/2}x\right] + D\cos\left[\frac{2\pi}{h}(2mE)^{1/2}x\right] \quad (3\text{-}37)$$

and, finally, in region III, V > E, x > L,

$$\Psi_{III} = F\exp\left\{\frac{2\pi}{h}[2m(V-E)]^{1/2}x\right\}$$

$$+ G\exp\left\{-\frac{2\pi}{h}[2m(V-E)]^{1/2}x\right\} \quad (3\text{-}38)$$

Now the criteria for meaningful solutions must be applied to these expressions for Ψ. First, the constraint that Ψ must be finite will simplify the situation:

- as $x \to -\infty$, B must $\to 0$, so that Ψ_I does not become infinite, and
- as $x \to +\infty$, F must $\to 0$, so that Ψ_{III} does not become infinite.

In addition, Ψ and $\partial\Psi/\partial x$ must be continuous. Thus, at the boundaries of region II:

at x = 0	at x = L
$\Psi_I = \Psi_{II}$	$\Psi_{II} = \Psi_{III}$
$\frac{\partial\Psi_I}{\partial x} = \frac{\partial\Psi_{II}}{\partial x}$	$\frac{\partial\Psi_{II}}{\partial x} = \frac{\partial\Psi_{III}}{\partial x}$

Taking the partial derivative of Ψ_{II} with respect to x

$$\frac{\partial\Psi_{II}}{\partial x} = C\frac{2\pi}{h}(2mE)^{1/2}\cos\left[\frac{2\pi}{h}(2mE)^{1/2}x\right]$$

$$-D\frac{2\pi}{h}(2mE)^{1/2}\sin\left[\frac{2\pi}{h}(2mE)^{1/2}x\right]$$

At x = 0, cos0 = 1 and sin0 = 0, so this becomes

$$\frac{\partial\Psi_{II}}{\partial x} = C\frac{2\pi}{h}(2mE)^{1/2}$$

Now the partial derivative of Ψ_I with respect to x is obtained, recalling that $B \to 0$,

$$\frac{\partial\Psi_I}{\partial x} = A\frac{2\pi}{h}[2m(v-E)]^{1/2}\exp\left\{\frac{2\pi}{h}[2m(V-E)]^{1/2}x\right\}$$

At x = 0 the exponential factor equals unity, so

$$\frac{\partial \Psi_I}{\partial x} = A\frac{2\pi}{h}\,[2m(V - E)]^{1/2}$$

In order that Ψ_I not be infinite, A must approach zero since V was postulated as being infinitely deep. Thus, Ψ_I must approach zero at the barrier (x = 0). This condition results in a negligibly small probability of finding the particle in region I because $\Psi_I{}^*\Psi_I \cong 0$.

In region III, Ψ_{III} also will approach zero because of the magnitude of the potential barrier, V. Continuity must be preserved at the boundaries of region II. Thus, at x = 0 and x = L, Ψ_{II} must approach zero, since both Ψ_I and Ψ_{III} do so. To meet this condition at x = 0, the coefficient D must equal zero, since cos 0 = 1. Thus, at x = 0,

$$\Psi_{II} = C\sin\left[\frac{2\pi}{h}(2mE)^{1/2}x\right]_{x=0} = 0$$

In order that Ψ_{II} equal zero at x = L,

$$C\sin\left[\frac{2\pi}{h}(2mE)^{1/2}x\right]_{x=L} = 0$$

But C does not equal zero, so, for this to be the case, it follows that

$$\frac{2\pi}{h}(2mE)^{1/2}L = n\pi$$

where n is an integer. Solving for E, the energy of the electron is given by

$$E = \frac{n^2h^2}{8mL^2} \tag{3-39}$$

Now the eigenvalue, E, can only take on certain allowed, integral values which are determined by the integer, n. This limitation arose from the restriction placed upon Ψ_{II}; it was contained within the well. When Ψ_{II} is made to approach zero at the boundaries of the well, and recalling that it is a periodic function of x, this means that it can only have an integral number of half wavelengths in the potential well over the interval between 0 and L. This number is determined by the value of the integer, n. This, of course, means that the energies (eigenvalues) will vary discretely with n.

It will be recalled that when the variables of Equation 3-29 were separated into two equations, Equations 3-30 and 3-32, it was noted that only certain values of E would satisfy the conditions. These are natural solutions to Schrödinger's equation and represent discrete energy states. These values are the eigenvalues. The Ψ functions which correspond to these eigenvalues are known as "eigenfunctions".

The eigenvalues thus arise naturally from solutions of the Schrödinger equation. This is in contrast to the Bohr approach wherein the angular momenta of electrons had to be postulated to be in integral multiples of $h/2\pi$ in order to ensure discrete energies. Thus, the application of the Schrödinger equation, based upon probability rather than classical mechanics, eliminates many of the difficulties encountered in the earlier model.

3.6. COMPARISON OF QUANTUM AND CLASSICAL MECHANICS

The eigenvalues are discrete since n is an integer, Equation 3-39. This results from

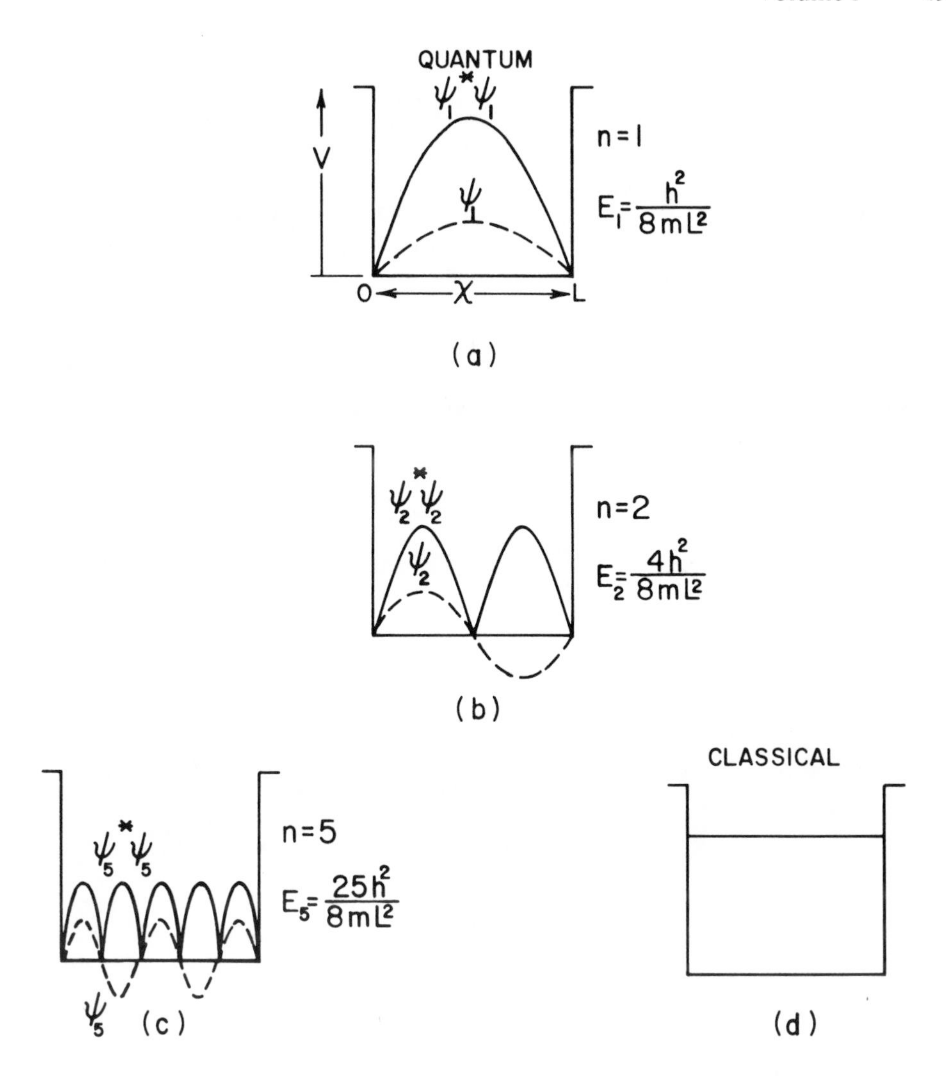

FIGURE 3-2. Probabilities of finding an electron in a potential well ($\Psi^*\Psi$ scales are compressed).

the boundary conditions imposed upon Ψ_{II}. Consequently, the lowest allowed energy is given by n = 1. The case n = 0 is forbidden because this is the equivalent of allowing the wave function to equal zero (Section 3.3). However, average energies, which only can be calculated from discrete values, may have any nonzero value. The constraints placed upon Ψ_{II} also affect the probability of finding an electron in the well; this varies periodically across the well.

The classical mechanics, in contrast, is based upon a continuous energy spectrum, and, since no conditions are placed upon it, the energy can equal zero. (Compare this with the behavior described in Section 4.2 in Chapter 4.) Another result of this is that the probability of finding the electron is the same at any position across the well.

The probabilities of finding the electron for both cases are shown in Figure 3-2.

In Figure 3-2a, where n = 1, just one-half of a wavelength of Ψ is contained in the well. The probability density of finding the electron, $\Psi^*\Psi$, varies from x = 0 to x = L. In Figure 3-2b, where one complete wavelength of Ψ is contained within the well, the probability varies in a more complicated way. The probability of finding an electron goes through zero in Figure 3-2b. How can this be? This question is not allowed, since, according to the Heisenberg Uncertainty Principle, when the exact position of the electron is known, none of its other properties can be known. For where $n = 5$, in Figure 3-2c, the probability of finding the electron becomes less variable. It thus can be seen that as n becomes sufficiently large, the probability of finding the electron in the well approaches the classical case, shown in Figure 3-2d, as a limit. Here, the probability is constant across the well. This behavior is in agreement with the Bohr correspondence principle, since the quantum-mechanic case approaches the classical case in the appropriate limit.

3.7. ELECTRON IN A FINITE WELL

The behavior of an electron in a deep, but not an infinite, one-dimensional well will now be considered. It was found earlier that

$$\Psi_I = A\exp\left\{\frac{2\pi}{h}\left[2m(V-E)\right]^{1/2}x\right\}; \quad x < 0$$

since it was shown that the coefficient B must approach zero. The wave function in the well was given as

$$\Psi_{II} = C\sin\left[\frac{2\pi}{h}(2mE)^{1/2}x\right] + D\cos\left[\frac{2\pi}{h}(2mE)^{1/2}x\right]; \quad 0 < x < L$$

Now evaluating these functions at $x = 0$, it is seen that $\Psi_I = A \exp[0]$ so that

$$\Psi_I = A$$

and since $\sin 0 = 0$ and $\cos 0 = 1$,

$$\Psi_{II} = D$$

The criteria for meaningful solutions are invoked again. Since continuity must exist across the boundary, $\Psi_I = \Psi_{II}$ and so $A = D$. Also, the derivatives must be continuous at the wall. So, at $x = 0$,

$$\frac{\partial \Psi_I}{\partial x} = A\frac{2\pi}{h}\left[2m(V-E)\right]^{1/2}$$

and

$$\frac{\partial \Psi_{II}}{\partial x} = C\frac{2\pi}{h}(2mE)^{1/2}$$

These derivatives must also be equal, thus

$$A\frac{2\pi}{h}\left[2m(V-E)\right]^{1/2} = C\frac{2\pi}{h}(2mE)^{1/2}$$

and, rearranging

$$A = \frac{C(2mE)^{1/2}}{\left[2m(V-E)\right]^{1/2}} = C\left[\frac{E}{V-E}\right]^{1/2}$$

It was just shown that A = D, so

$$D = C\left[\frac{E}{V - E}\right]^{1/2}$$

This will be used shortly.

Continuity must also exist at x = L; that is, $\Psi_{II} = \Psi_{III}$,

$$C\sin\left[\frac{2\pi}{h}(2mE)^{1/2}L\right] + D\cos\left[\frac{2\pi}{h}(2mE)^{1/2}L\right]$$

$$= G\exp\left\{-\frac{2\pi}{h}\left[2m(V - E)\right]^{1/2}L\right\}$$

Now, for convenience, let the parameters

$$\beta = \frac{2\pi}{h}(2mE)^{1/2}$$

and

$$\alpha = \frac{2\pi}{h}\left[2m(V - E)\right]^{1/2}$$

From these it will be noted that, for later use,

$$\frac{\beta}{\alpha} = \left[\frac{E}{V - E}\right]^{1/2}$$

Now, the equation representing $\Psi_{II} = \Psi_{III}$, using the above parameters, becomes

$$C\sin\beta L + D\cos\beta L = Ge^{-\alpha L}$$

The derivatives of these functions also must be equal. Thus,

$$\beta C\cos\beta L - \beta D\sin\beta L = -\alpha Ge^{-\alpha L}$$

This expression is now divided by the original equation to obtain

$$\frac{\beta C\cos\beta L - \beta D\sin\beta L}{C\sin\beta L + D\cos\beta L} = -\frac{\alpha Ge^{-\alpha L}}{Ge^{-\alpha L}}$$

Factoring β and dividing both sides of this equation by β gives

$$\frac{C\cos\beta L - D\sin\beta L}{C\sin\beta L + D\cos\beta L} = -\frac{\alpha}{\beta}$$

Multiplying by (−1) and rearranging results in

$$\frac{D\sin\beta L - C\cos\beta L}{D\cos\beta L + C\sin\beta L} = \frac{\alpha}{\beta}$$

Using the relationship previously found between D and C, the following is obtained

$$\frac{C\left[\frac{E}{V-E}\right]^{1/2}\sin\beta L - C\cos\beta L}{C\left[\frac{E}{V-E}\right]^{1/2}\cos\beta L + C\sin\beta L} = \frac{\alpha}{\beta}$$

The factor C vanishes from the fraction. Then, recalling the relationship previously found between α and β:

$$\frac{\frac{\beta}{\alpha}\sin\beta L - \cos\beta L}{\frac{\beta}{\alpha}\cos\beta L + \sin\beta L} = \frac{\alpha}{\beta}$$

Clearing α and rearranging the terms gives

$$\beta^2\sin\beta L - \alpha\beta\cos\beta L = \alpha\beta\cos\beta L + \alpha^2\sin\beta L$$

Collecting terms

$$(\beta^2 - \alpha^2)\sin\beta L = 2\alpha\beta\cos\beta L$$

or

$$\tan\beta L = \frac{\sin\beta L}{\cos\beta L} = \frac{2\,\alpha\beta}{\beta^2 - \alpha^2}$$

Finally, substituting the values of the parameters of α and β, and simplifying, gives the relationship being sought:

$$\tan\left[\frac{2\pi}{h}(2mE)^{1/2}L\right] = \frac{2\left[E(V-E)\right]^{1/2}}{2E - V}$$

This expression cannot be solved directly for E. It can be solved graphically. However, an approximate solution will be given here. Consider the right-hand side of this equation and recall that $V \gg E$, then, dividing numerator and denominator by V,

$$\frac{2\left[E(V-E)\right]^{1/2}}{2E - V} = \frac{2\left[\frac{E}{V}\left(1-\frac{E}{V}\right)\right]^{1/2}}{2\frac{E}{V} - 1} \approx \frac{2\left[\frac{E}{V}(1-0)\right]^{1/2}}{2\times 0 - 1} \approx -2\left[\frac{E}{V}\right]^{1/2} \approx 0$$

Thus, it may be approximated that

$$\tan\left[\frac{2\pi}{h}L(2mE)^{1/2}\right] \approx 0$$

and, for this to be so, the following must be true:

$$\frac{2\pi}{h}L(2mE)^{1/2} = n\,\pi$$

Squaring both sides of this equation, and noting that π vanishes,

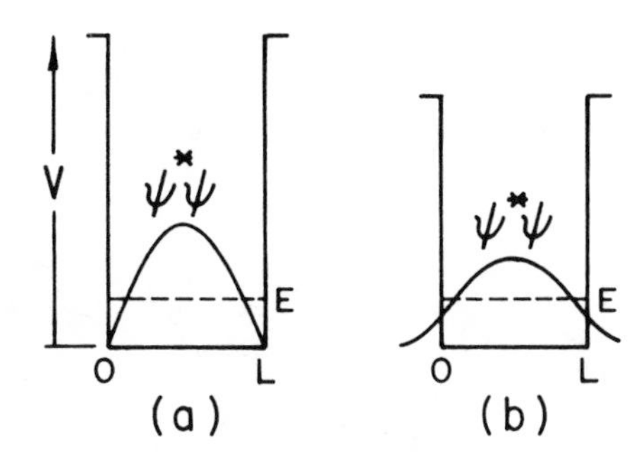

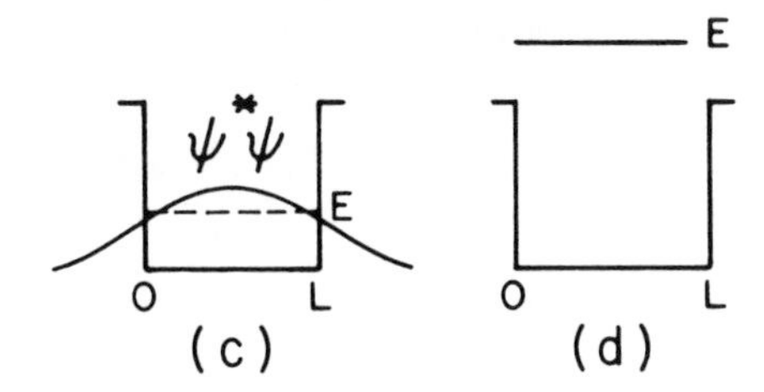

FIGURE 3-3. The effect of well depth upon the probability of finding an electron.

$$\frac{4L^2}{h^2} \cdot 2mE = n^2$$

then, solving for E

$$E = \frac{n^2h^2}{8mL^2} \qquad (3\text{-}40)$$

This is the same result as that obtained for a much higher barrier (see Equation 3-39). Discrete energy levels are obtained again.

Now, consider the effect of further decreasing the depth of the potential well. The same equations used for the three regions are used. However, these are no longer made to equal zero at x = 0 and x = L. Thus Ψ_I and Ψ_{III} drop off exponentially into regions I and III. The closer V approaches E, the greater are the values of Ψ_I and Ψ_{III} at x = 0 and x = L, respectively, and they drop off less rapidly. At the point where E > V, the particle is no longer affected by the potential. This removes all of the restrictions upon E, and the classical case of the unrestricted, or free, particle prevails. These effects are shown schematically in Figure 3-3.

It should also be noted that when x = L is small, of atomic dimensions, the energy levels are relatively widely separated. However, as L gets larger and normal crystal dimensions are approached, the energy levels become closer and closer together until a quasi-continuum is approached, in agreement with the Bohr correspondence principle. This is shown schematically in Figure 3-4.

The case where many electrons are in a well of crystal dimensions, such as that shown in Figure 3-4B, has many important applications which are discussed in Chapter 5. Some of these include the Sommerfeld theory, the Fermi-Dirac statistics, and electron transport processes.

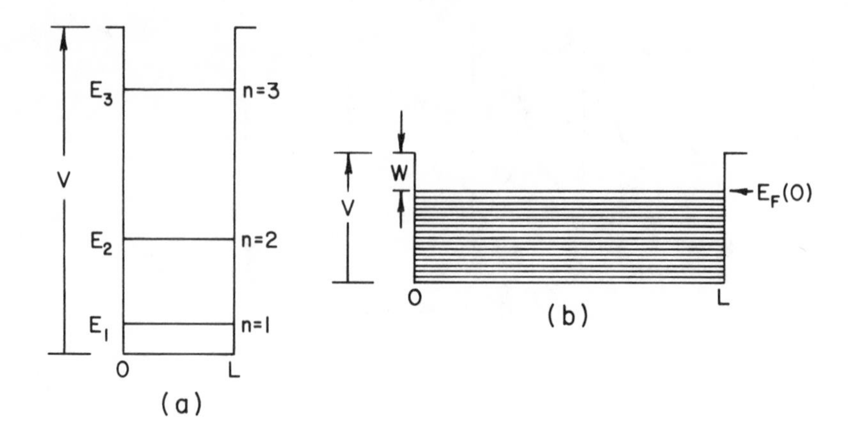

FIGURE 3-4. The effect of well size upon energy states. (a) For small L the states are widely separated; (b) For large L the narrowly separated states approach a quasi-continuum. See Figure 1-2 in Chapter 1, Figures 5-5 and 5-20 in Chapter 5, and Equation 5-29 in Chapter 5. (Figure 3-4b after C. W. Curtis. With permission.)

3.8. THREE-DIMENSIONAL POTENTIAL WELL

Up to this point, solutions to Schrödinger's equation have been found only for one-dimensional wells. Consideration will now be given to the behavior of an electron in a three-dimensional box with infinitely high walls, as shown schematically in Figure 3-5. The variation of the potential along the x−z and y−z planes is the same as for the one-dimensional case (Section 3.5), namely

- $V(x,y,z) = 0$ in the box
- $V(x,y,z) = \infty$ outside the box

Since V(x,y,z) is zero within the box, Schrödinger's time-independent equation becomes

$$\frac{\partial^2 \Psi}{\partial x^2} + \frac{\partial^2 \Psi}{\partial y^2} + \frac{\partial^2 \Psi}{\partial z^2} + \frac{8\pi^2 m}{h^2} E\Psi = 0 \qquad (3\text{-}41)$$

where Ψ is a function of x, y, and z. It is helpful to separate these variables, so solutions will be sought of the form,

$$\Psi = \Psi(x,y,z) = X(x)Y(y)Z(z) \qquad (3\text{-}42)$$

in which each of the individual expressions are functions of a single variable. Such separable solutions do not always exist; they depend upon the form of the potential energy. The second derivatives of Equation 3-42 are found to be

$$\frac{\partial^2 \Psi(x,y,z)}{\partial x^2} = Y(y)Z(z)\frac{\partial^2 X(x)}{\partial x^2} \qquad (3\text{-}43a)$$

$$\frac{\partial^2 \Psi(x,y,z)}{\partial y^2} = X(x)Z(z)\frac{\partial^2 Y(y)}{\partial y^2} \qquad (3\text{-}43b)$$

$$\frac{\partial^2 \Psi(x,y,z)}{\partial z^2} = X(x)Y(y)\frac{\partial^2 Z(z)}{\partial z^2} \qquad (3\text{-}43c)$$

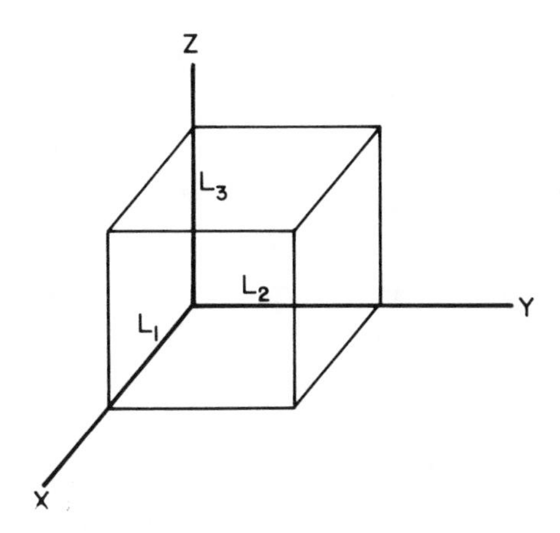

FIGURE 3-5. Sketch of a three-dimensional well.

By means of Equations 3-43a, 3-43b, and 3-43c, Equation 3-41 becomes

$$Y(y)Z(z)\frac{\partial^2 X(x)}{\partial x^2} + X(x)Z(z)\frac{\partial^2 Y(y)}{\partial y^2} + X(x)Y(y)\frac{\partial^2 Z(z)}{\partial z^2} + \frac{8\pi^2 m}{h^2} EX(x)Y(y)Z(z) = 0 \qquad (3\text{-}44)$$

When Equation 3-44 is divided through by X(x)Y(y)Z(z),

$$\frac{1}{X(x)}\frac{\partial^2 X(x)}{\partial x^2} + \frac{1}{Y(y)}\frac{\partial^2 Y(y)}{\partial y^2} + \frac{1}{Z(z)}\frac{\partial^2 Z(z)}{\partial z^2} + \frac{8\pi^2 m}{h^2} E = 0 \qquad (3\text{-}45)$$

Each of three terms in Equation 3-45 is a function of a single, different variable. The sum of these three terms is a constant. This can only be true if each of the terms is equal to a constant, or eigenvalue. Therefore,

$$\frac{1}{X(x)}\frac{\partial^2 X(x)}{\partial x^2} = -\frac{8\pi^2 m}{h^2} E_x \qquad (3\text{-}46a)$$

$$\frac{1}{Y(y)}\frac{\partial^2 Y(y)}{\partial y^2} = -\frac{8\pi^2 m}{h^2} E_y \qquad (3\text{-}46b)$$

and

$$\frac{1}{Z(z)}\frac{\partial^2 Z(z)}{\partial z^2} = -\frac{8\pi^2 m}{h^2} E_z \qquad (3\text{-}46c)$$

where the eigenvalues E_x, E_y, and E_z are constants and

$$E_x + E_y + E_z = E \qquad (3\text{-}46d)$$

These equations are treated individually as shown by using Equation 3-46a as an example. Upon rearrangement they take the form:

$$\frac{\partial^2 X(x)}{\partial x^2} + \frac{8\pi^2 m}{h^2} E_x X(x) = 0$$

These equations are of the same form as that previously obtained for the one-dimensional well (Section 3.5). The application of the same criteria and boundary conditions as were previously used gives the following:

$$E_x = \frac{n_x^2 h^2}{8mL_x^2} \quad (3\text{-}47a)$$

$$E_y = \frac{n_y^2 h^2}{8mL_y^2} \quad (3\text{-}47b)$$

$$E_z = \frac{n_z^2 h^2}{8mL_z^2} \quad (3\text{-}47c)$$

In which n_x, n_y, and n_z are integers. Now making use of Equation 3-46d,

$$E = E_x + E_y + E_z = \frac{h^2}{8m}\left[\frac{n_x^2}{L_x^2} + \frac{n_y^2}{L_y^2} + \frac{n_z^2}{L_z^2}\right] \quad (3\text{-}48)$$

If the box is a cube, $L_x = L_y = L_z = L$, and

$$E = \frac{h^2}{8mL^2}\left[n_x^2 + n_y^2 + n_z^2\right] \quad (3\text{-}49)$$

The three integers n_x, n_y, and n_z are the three quantum numbers required to specify each state in rectangular coordinates. Again, none of these may be equated to zero, since this would be the same as setting the wave function equal to zero; it will be recalled (Section 3.3) that this violates the criteria for such wave functions. Equations 3-48 and 3-49 give the probable energy of an electron in terms of its most likely position. A portion of this energy has been associated with each coordinate, the total energy being the sum of these.

Frequent use will be made of Equation 3-49 in explaining the properties of solids.

3.9. THE PAULI EXCLUSION PRINCIPLE

The preceding derivations and illustrations have considered only one electron in potential wells. It might be thought that as more electrons are added all of these would be found at some common, lowest energy level. If this were so, all of the elements would have very similar properties. However, this is known not to be the case. The work begun by Mendeleev on the Periodic Table shows that the elements have widely differing chemical behavior which varies uniformly and periodically with their atomic number. The work on light spectra (Chapter 1) also shows that the electrons must occupy various different energy levels. It has also been shown that the discrete X-ray spectrum of an element is characteristic of that element. Since this involves the inner electrons which normally do not enter into bonding or chemical reactions, these too must be in different energy states.

In considering the presence of more than one electron in a potential well, such as that constituted by a positively charged nucleus, it is natural to think of the repulsions of the similarly charged electrons and their attractions to the nucleus. The behavior of each one of the electrons would affect the behavior of all of the others. In a sense this mutual interaction could be thought of as the sole basis for difference in atomic behaviors. However, such a model based only upon coulombic attractions could not account for the phenomena noted above.

It will be recalled that the three-body problem (sun, earth, and moon, or a nucleus and two electrons) cannot be solved exactly. The complications arising in the consideration of atoms with atomic numbers of the order of 100 can be understood readily. This difficulty may be avoided. As has been shown in this chapter, and in Chapter 1, the electron behaviors can be described readily in terms of energies and quantum numbers and not in terms of their mutual electrostatic behaviors. A way of doing this is by the use of the integers used in Equation 3-48. If additional electrons are successively added to a nucleus, or a well, the energy of the system and their quantum numbers must change.

Based upon the behavior of spectra, Pauli (1925) stated the exclusion principle which describes this behavior. This postulates that no two electrons can be in the same quantum state. In terms of Equation 3-49, this means that no two electrons can have the same integers n_x, n_y, n_z in the same order. This description is incomplete, since a fourth quantum number, spin, must also be included. The wave mechanics explaining this principle are outlined in Section 3.11.

When the Exclusion Principle is applied to the electrons about a nucleus, it results in electron configurations which are in agreement with the behaviors previously noted: the periodic chemical behavior and light and X-ray spectra, to mention only a few phenomena, are explained readily.

However, electrons with different quantum numbers can have the same energies. As examples, electrons with the following quantum numbers (5,1,1), (1,5,1), (1,1,5), and (3,3,3) all have different quantum numbers, but have the same energy: $27h^2/8mL^2$, according to Equation 3-49. These are known as degenerate states. A nondegenerate electron state is one for which only one wave function exists corresponding to the given energy. The state (1,1,1) is nondegenerate; no other state can have $E = 3h^2/8mL^2$.

In a well of crystal dimensions, such as shown in Figure 3-4b, the order of 10^{22} or 10^{23} valence electrons may be involved. Obviously, this crystal will contain many electrons in degenerate states. The large number of electron energy levels "packed" into a relatively small range of energy results in the energy levels being very close together but still discrete. This condition is called a quasi-continuum and it approaches the classical idea of a continuous range of energies as a limit. This too conforms to the Bohr Correspondence Principle (see Figures 1-2 in Chapter 1 and 5-5 in Chapter 5).

3.10. HARMONIC OSCILLATORS

Unlike electrons, each ion within a solid is relatively restricted in its motion. These ions oscillate in a simple harmonic mode about a mean, or equilibrium, position. The problem of determining the probability of finding such a particle will be approached both from the classical and quantum mechanical methods. This will provide an additional application of Schrödinger's equation and a basis for understanding lattice vibrations which will be of use in explaining the thermal properties of solids (Chapter 4).

First, consider the classical case as sketched in Figure 3-6. The particle of mass, m, is acted upon by a potential $V(x) = 1/2\ Kx^2$ and a restoring force $F = -Kx$. The classical solution starts with

$$F = ma = m\frac{d^2x}{dt^2}$$

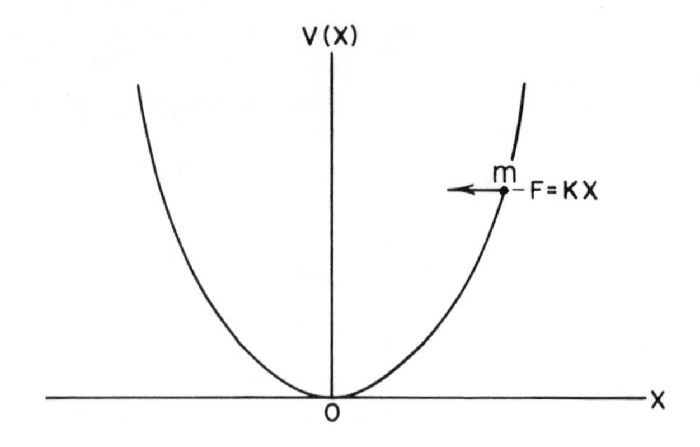

FIGURE 3-6. Harmonic oscillator.

and equating the two expressions for F gives

$$-Kx = m\frac{d^2x}{dt^2} \tag{3-50a}$$

or

$$\frac{d^2x}{dt^2} + \frac{Kx}{m} = 0 \tag{3-50b}$$

The equation for the position, x, of the oscillating particle is given by

$$x = x_m \cos(2\pi\nu t) \tag{3-51}$$

where x_m is maximum displacement and ν is its frequency. The first and second derivatives of Equation 3-51 are

$$dx = x_m(-2\pi\nu)\sin(2\pi\nu t)dt \tag{3-52a}$$

and

$$d^2x = x_m(-4\pi^2\nu^2)\cos(2\pi\nu t)dt^2 \tag{3-52b}$$

At $t = 0$, $x = x_m$ and, since $\cos 0 = 1$,

$$\frac{d^2x}{dt^2} = -4\pi^2\nu^2 x_m \tag{3-53}$$

Equation 3-53 can be used with Equation 3-50a to determine K

$$\frac{d^2x}{dt^2} = -\frac{Kx}{m} = -4\pi^2\nu^2 x \tag{3-54}$$

and it is found that

$$K = 4\pi^2\nu^2 m \tag{3-55}$$

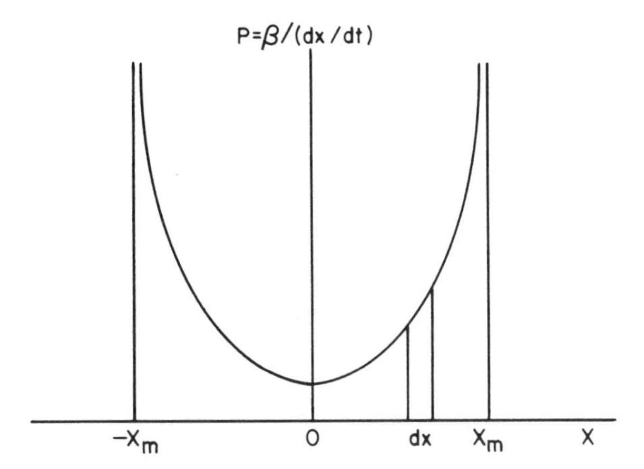

FIGURE 3-7. Probable position of a classical harmonic oscillator. (The probability scale is expanded.)

The potential energy may be reexpressed from this from

$$V(x) = \frac{1}{2} Kx_m^2 = \frac{1}{2}(4\pi^2\nu^2 m)x_m^2 \tag{3-56}$$

Solving Equation 3-56 for x_m gives the more general expression

$$x_m = \pm\frac{1}{\pi\nu}\left[\frac{V(x)}{2m}\right]^{1/2} \tag{3-57}$$

This is the classical limit for the position of the oscillating particle. On a classical basis there is zero probability of finding the particle beyond $\pm\ x_m$. This will now be explained.

The classical probability, P, of finding the particle in the region dx, (Figure 3-7), is proportional to the time spent by the particle in that region, or

$$P\,dx = \beta\,dt \tag{3-58a}$$

where β is a constant of proportionality. Or,

$$P = \beta\frac{dt}{dx} = \beta\frac{1}{v} \tag{3-58b}$$

where v is the velocity of the particle. However, as x approaches x_m, v slowly approaches zero. Thus, P must approach infinity. Since v approaches zero slowly, P approaches infinity slowly; the area under the curve is finite. Since P becomes very large as x approaches x_m, there is zero probability of finding the particle beyond x_m. This verifies the prior statement that when treated classically, the particle always will be found within the potential well.

Now consider the quantum-mechanic approach to this problem. Starting with Equation 3-34 and the expression for V(x), Equation 3-56,

$$\frac{\partial^2\Psi(x)}{\partial x^2} + \frac{8\pi^2 m}{h^2}\left[E - 2\pi^2\nu^2 mx^2\right]\Psi(x) = 0 \tag{3-59}$$

Now consider the wave function, where $\alpha = 2\pi^2 m\nu/h$,

$$\Psi_o(x) = \exp(-\alpha x^2) \tag{3-60}$$

as a trial solution to Equation 3-59. Its derivatives are

$$\frac{\partial \Psi_o(x)}{\partial x} = -2\alpha x \exp(-\alpha x^2)$$

and

$$\frac{\partial^2 \Psi_o(x)}{\partial x^2} = -2\alpha \exp(-\alpha x^2) + 4\alpha^2 x^2 \exp(-\alpha x^2) \tag{3-61}$$

or

$$\frac{\partial^2 \Psi_o(x)}{\partial x^2} = -2\alpha \Psi_o(x) + 4\alpha^2 x^2 \Psi_o(x)$$

$$= -(2\alpha - 4\alpha^2 x^2)\Psi_o(x) \tag{3-62}$$

Now substituting for α

$$\frac{\partial^2 \Psi_o(x)}{\partial x^2} = -\left[2\frac{2\pi^2 m\nu}{h} - 4\frac{4\pi^4 m^2 \nu^2}{h^2} x^2\right] \Psi_o(x) \tag{3-63}$$

$$\frac{\partial^2 \Psi_o(x)}{\partial x^2} = -\frac{4\pi^2 m}{h^2} [h\nu - 4\pi^2 m\nu^2 x^2]\, \Psi_o(x) \tag{3-64}$$

$$\frac{\partial^2 \Psi_o(x)}{\partial x^2} = -\frac{8\pi^2 m}{h^2}\left[\frac{h\nu}{2} - 2\pi^2 m\nu^2 x^2\right] \Psi_o(x) \tag{3-65}$$

Or,

$$\frac{\partial^2 \Psi_o(x)}{\partial x^2} + \frac{8\pi^2 m}{h^2}\left[\frac{h\nu}{2} - 2\pi^2 m\nu^2 x^2\right] \Psi_o(x) = 0 \tag{3-66}$$

Equation 3-66 is identical to Equation 3-59 and $\Psi_o(x)$ is a solution to Equation 3-59 if

$$E = E_o = \frac{h\nu}{2} \tag{3-67}$$

Equations 3-59 and 3-66 are now the same and are identical to the Schrödinger time-independent equation. Thus, E_o is an eigenvalue and $\Psi_o(x)$ is an eigenfunction of Schrödinger's equation.

In a similar way, using the wave function,

$$\Psi_1(x) = x\exp(-\alpha x^2) \tag{3-68}$$

it can be shown that

$$E_1 = \frac{3}{2} h\nu \tag{3-69}$$

A third eigenfunction

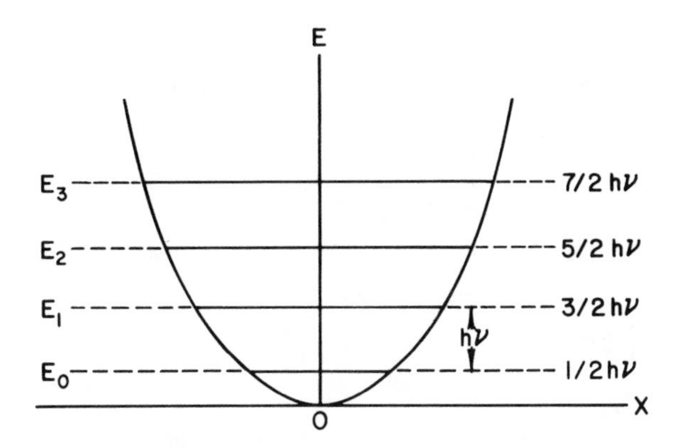

FIGURE 3-8. Quantized energy levels of a simply oscillating particle.

$$\Psi_2(x) = (1 - 4\alpha x^2)\exp(-\alpha x^2) \tag{3-70}$$

will give an eigenvalue

$$E_2 = \frac{5}{2} h\nu \tag{3-71}$$

The selection of the eigenfunctions, Equations 3-61, 3-68, and 3-70, is not arbitrary but may be found by direct methods. These methods are beyond the scope of this text. The eigenvalues found thus far may be summarized as follows

$$E_0 = \frac{1}{2} h\nu = \left(0 + \frac{1}{2}\right) h\nu$$

$$E_1 = \frac{3}{2} h\nu = \left(1 + \frac{1}{2}\right) h\nu$$

$$E_2 = \frac{5}{2} h\nu = \left(2 + \frac{1}{2}\right) h\nu$$

or, generalizing,

$$E_n = \left(n + \frac{1}{2}\right) h\nu \tag{3-72}$$

Here, n may take on the value of zero to give the lowest energy of the oscillating particle. This condition, $n = 0$, will be referred to as the zero-point energy in Chapter 4. It will be noted that the oscillating particle can take on only energy values which differ from each other by exactly $h\nu$. This is shown in Figure 3-8. These ideas also will be helpful in the study of lattice vibrations.

The eigenfunctions will now be examined. These are shown in Figure 3-9. First consider the extent of these functions. Note that the "tails" extend beyond $\pm x_m$. This condition could not exist in the classical treatment, Equation 3-58b. It means that a small, real probability exists of finding the particle between $\pm x_m$ and $\pm$ infinity. As n becomes considerably larger, the tails shrink.

The wave function corresponding to the lowest energy state, the zero point energy, or ground state, E_o, is represented by Ψ_o. There is one maximum present in $\Psi_o^*\Psi_o$. The

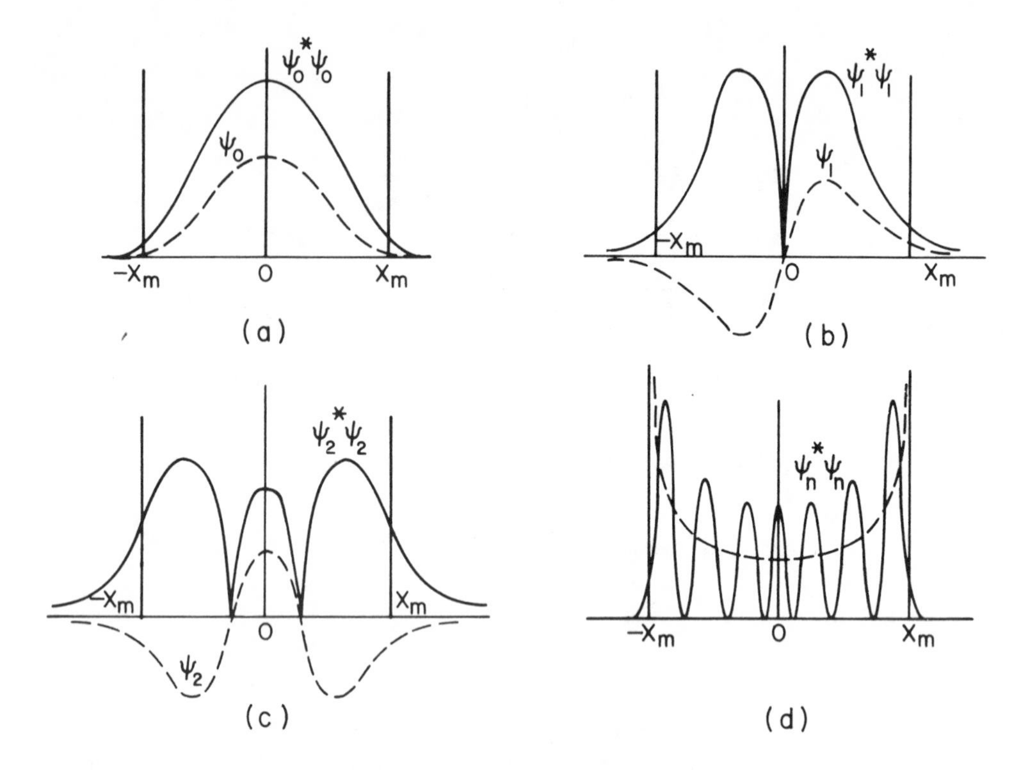

FIGURE 3-9. Normalized wave functions and probability densities for a simple harmonic oscillator. Scales of probability densities are compressed.

eigenvalue E_1 is the first excited state; the probability density $\Psi_1{}^*\Psi_1$ has two maxima. The second excited state has the eigenvalue E_2 obtained from Ψ_2, and $\Psi_2{}^*\Psi_2$ has three maxima. The probability density $\Psi_n{}^*\Psi_n$ has n + 1 maxima. Also, as n becomes sufficiently large, $\Psi_n{}^*\Psi_n$ approaches the classical case as the limit, in agreement with the correspondence principle. This is shown schematically in Figure 3-9d, where the broken curve indicates the average probability.

It is important to emphasize that discrete energy states for oscillating particles no longer need to be assumed as was done in order to explain black-body radiation. The discrete eigenvalues derive naturally from the solutions of Schrödinger's equation.

The concept of oscillating particles in discrete energy states is important in the consideration of the thermal properties of solids as in Chapter 4.

3.11. APPLICATION TO ATOMIC STRUCTURE

Previous sections have considered the Schrödinger model of the behavior of individual electrons as affected by potential wells. The wave-mechanic approach is extended in this section to provide an outline of the interactions of electrons with nuclei.

The simplest atom, hydrogen, is used as the basis for understanding the electron configurations of atoms. It consists of a positive nucleus, a single proton, with one electron in "orbit" around it. Instead of a cubic well, as in Section 3.7, the potential is spherically symmetrical about the nucleus. Since this potential is only a function of the distance (r) between the charged particles, the spherical potential well is described more accurately by modifying Equation 1-20 to include the dielectric constant of vacuum, ε_o, to give

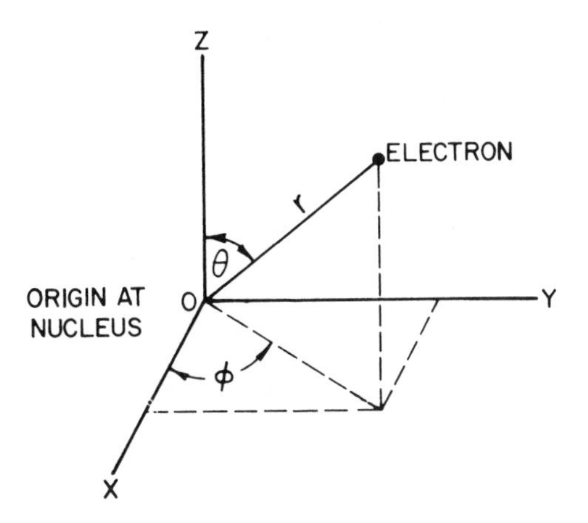

FIGURE 3-10. Spherical coordinates for a hydrogen atom.

$$V(r) = -\frac{e^2}{4\pi\epsilon_0 r} \tag{3-73}$$

when the origin is taken at the nucleus, which is considered to be point mass. This system is shown in Figure 3-10.

Here

$$x = r \sin\theta \cos\varphi$$

$$y = r \sin\theta \sin\varphi$$

$$z = r \cos\theta$$

and

$$r^2 = x^2 + y^2 + z^2 \tag{3-74a}$$

The wave equation is now a function of the variables r, θ, and φ and corresponds to Equation 3-34:

$$\frac{1}{r^2}\frac{\partial}{\partial r}\left[r^2\frac{\partial\Psi}{\partial r}\right] + \frac{1}{r^2\sin\theta}\frac{\partial}{\partial\theta}\left[\sin\theta\frac{\partial\Psi}{\partial\theta}\right] + \frac{1}{r^2\sin^2\theta}\frac{\partial^2\Psi}{\partial\varphi^2}$$

$$+ \frac{8\pi^2}{h^2}[E - V(r)]\Psi = 0 \tag{3-75}$$

Solutions are sought which are of the form

$$\Psi(r,\theta,\varphi) = R(r)\Theta(\theta)\Phi(\varphi) \tag{3-76}$$

in a way analogous to the treatment of Equation 3-28.

Separations are performed and three ordinary, linear differential equations are obtained; their solutions provide three quantum numbers. The first of these is the principle quantum number, n. The second is the angular momentum, or azimuthal, quantum number, ℓ. The third is m_ℓ, the magnetic quantum number. The quantum

Table 3-1
DESIGNATION OF ELECTRON STATES

n	Maximum (ℓ)	Maximum number of states	Designation of states
1	0	2	s
2	1	6	p
3	2	10	d
4	3	14	f

numbers n and ℓ are related to R(r), ℓ and m_ℓ are related to Θ (θ) and m_ℓ is related to ϕ (φ).

The principle quantum number, n, is a positive integer. In terms of the Bohr model, it specifies the major axis of an elliptical orbit. It denotes the probable size of the "orbit" and the energy shell (see Section 1.5 in Chapter 1). As n increases, the probability of finding the electron moves away from the nucleus. The letters K, L, M, N, etc., corresponding to n = 1, 2, 3, 4, etc., respectively, are used in X-ray and spectroscopic work. The total possible number of electrons with a given value of n is $2n^2$. As previously noted, the gross energy level is determined by n.

The second quantum number (ℓ) is a measure of the angular momentum of the orbital motion of the electron. In the Bohr representation, ℓ also fixes the minor axes of the "orbits". It can assume integral values from zero to (n − 1), each of which denotes small differences in energy. For a given value of ℓ the maximum number of states is given by $2(2\ell + 1)$. These relationships are given in Table 3-1.

The designations used in Table 3-1 are the same as those given in Section 1.4 in Chapter 1. As the numerical value of ℓ increases, the probability of finding the electron at slightly greater distances from the nucleus increases. As such, ℓ represents small energy variations in the sublevels within an energy level given by a particular n. The energy differences resulting from electron transitions in ℓ within a given n are smaller compared to electron transitions between levels.

Where the wave functions are such that $\ell = \ell_{max}$, the electronic transitions from n to n − 1 results in a more circular "orbit". Where $\ell < \ell_{max}$, the "orbits" become more elliptical for increasing n. At $\ell = 0$ a linear wave function exists; here the probability of finding the electron passes through the nucleus. This state is spherically symmetrical since all spatial orientations of the linear function are equally probable. Some probability densities are shown in Figure 3.11.

The third quantum number m_ℓ also explains observed spectral behavior and is operative only in the presence of a magnetic field. It determines the spatial orientation of the plane of the "orbit". In the presence of a magnetic field m_ℓ can vary in integral steps from $-\ell$ to ℓ, including zero. Here, different values of m_ℓ become apparent from small variations in the energy of an atom depending upon its orientation in the magnetic field. This explains earlier experimental observations that under these conditions multiple spectral lines will appear. These lines represent wavelengths which are very close to that of the single line which appears in the absence of the magnetic field. This is one of the bases for the association of a magnetic moment with an electron. Such line splitting is known as the Zeeman effect. In the absence of a magnetic field, m_ℓ is unaffected and degenerate states result.

Equation 3-76 only provides three quantum numbers. An additional quantum number was needed to explain the observed spectral behavior of atoms in nonhomogeneous magnetic fields. Here, spectral lines of atoms with complete inner shells and one outer s electron ($\ell = 0$) were expected to be unaffected by the field. It was found, however, that some spectral lines of such atoms split into two lines. This behavior was originally

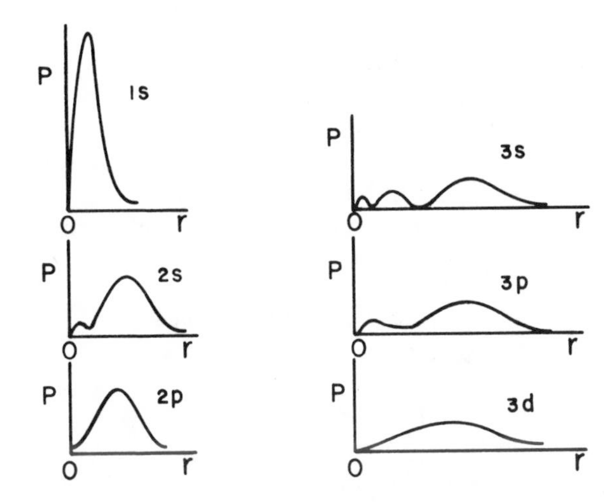

FIGURE 3-11. Probability densities of electron states of hydrogen (diagrams to same scale). (From Mahan, B. H., *College Chemistry,* Addison-Wesley, Reading, Mass., 1969, 379. With permission.)

explained on the basis that these electrons were considered to be spinning about their own axes while in "orbit" about the nucleus. This meant that two spin orientations must exist. In one of these orientations the electron can be thought to spin about its own axis in the same direction in which it is orbiting the nucleus; this is sometimes called "parallel spin". The other spin direction can be considered as an intrinsic electron rotation in a direction opposite to its orbital direction, or antiparallel spin. This behavior is a lower energy state than the parallel spin. For two such spin orientations the spin quantum number m_s is either $\pm$ 1/2. In actuality, spin cannot be described. The properties of electrons are such that they behave as though they spin.

See Section 8.3.2 in Chapter 8 for further details regarding ℓ, m_ℓ and m_s.

Thus, the concept of spin also arose from the necessity to explain observed spectral data. It was shown later by Dirac that electrons must possess both spins and angular momenta.

Reference has been made to the Bohr atom to help explain the quantum numbers obtained from Schrödinger's equation. The properties of an electron in an "orbit" about a nucleus are easy to picture using this model. However, when wave functions are used, probability distributions are obtained for the properties of an electron "orbiting" about the nucleus. It will be recognized from prior discussions that it is not possible to know these properties in the same sense as in the Bohr atom. It is thus necessary to expect on an intuitive basis that some type of electron orbital behavior must take place which includes momentum, magnetic moment, and spin. This must be the case because the experimentally observed phenomena can be explained only by the inclusion of these factors.

The solution just outlined for hydrogen is that for a two-body problem: one electron and a nucleus (proton). Where many electrons are involved such solutions become much more difficult, primarily because V(r) (Equation 3-73) becomes very complicated. This complexity arises from the necessity to include the combined effects of the coulombic attraction between the nucleus and each electron as well as all of the electron-electron interactions.

One way of dealing with this problem is known as the perturbation method. Because of the large disparity in the mass of an electron compared to that of a nucleus, especially as the atomic weight increases, the nucleus may be approximated as being at rest. This permits the use of the mass of the electron in Schrödinger's equation, rather than its effective mass, and simplifies matters; only relatively small errors are introduced. Using the helium atom as the least complicated means to describe this method, V(r) contains a symmetrical term for the attraction between the nucleus and each of the electrons and a third term giving the potential of the electron-electron influence. The third term is known as the perturbation when V(r) so constituted is used in Schrödinger's equation. It gets its name because exact solutions cannot be obtained when this term is present. This results from the fact that V(r) containing this term lacks complete spherical symmetry.

If the interaction term in V(r) is neglected, a zero-order approximation of the helium atom can be obtained; two wave functions which are essentially the same as that for the hydrogen atom result. This is the same as considering the two electrons to be independent of each other, a condition to be expected since the interaction term was dropped.

A much closer approximation of the helium atom may be made starting with Equation 3-21, where Ψ now is a function of x, y, and z. The Hamiltonian becomes $H = H_o + H_1$, the wave function becomes $\Psi = \Psi_o + \Psi_1$, and the eigenvalue becomes $E = E_o + E_1$. The terms with the zero subscript are essentially the same as those for the hydrogen atom and those with the unit subscript represent perturbations to approximate the electron-electron interactions. Thus

$$H\Psi = E\Psi$$

becomes

$$(H_O + H_1)(\Psi_O + \Psi_1) = (E_O + E_1)(\Psi_O + \Psi_1)$$

The terms similar to those for the hydrogen atom and those containing the perturbation terms are collected and retained, while those involving products of perturbation terms are neglected. A first-order approximation must follow because of this. Solutions for ψ_1 are obtained in terms of the known Ψ_o. These provide eigenvalues which agree with experimental data within about 5%. This approach is for the nondegenerate case in which spin has not been considered.

However, degeneracy of orbital functions is present in most atoms and must be included. Linear combinations of degenerate wave functions, each corresponding to a given eigenvalue, are employed using the same procedure as that outlined for the nondegenerate case, spin again being omitted. The number of roots obtained from the expression for the perturbation energy gives the degree of degeneracy of the atom.

Implicit in both of the above approximations is that the wave functions and perturbation functions are symmetric. Thus, for a helium atom,

$$E_1 = H_{11} + H_{12}$$

where the second subscript denotes the electron. Where the condition

$$E_1 = H_{11} - H_{12}$$

exists, the wave functions are termed antisymmetric. This change in sign comes about when the two electrons are interchanged. The change in the spatial coordinates of the electrons results in antisymmetric wave functions. This concept is an important factor in the Pauli exclusion principle. Both symmetric and antisymmetric wave functions also will be employed as a basis for the derivation of band theory (see Section 5.7 in Chapter 5).

The final factor to be considered is spin. This property was not included in the foregoing description for helium because it is not derived from Schrödinger's equation. The inclusion of this factor increases the number and complexity of the resulting wave functions; these are known as spin-orbital wave functions. The spin-orbital functions are symmetric if both components are symmetric; they are antisymmetric if one component is antisymmetric. Any atom with two or more electrons is postulated to be antisymmetric. Experimental observations show this to be true for helium. With this postulate as a basis, an exchange of two electrons will change the sign of the wave function. This is the quantum mechanical basis for the Pauli exclusion principle, since it explains why no two electrons can have the same set of quantum numbers.

The solution of Schrödinger's equation for the hydrogen atom, with one electron in a spherical potential, was used to obtain three quantum numbers to which a fourth, spin, was added. The theoretical basis for the Pauli exclusion principle, governing allowed values of the quantum numbers, was explained using helium, a two-electron atom. A justification for the extension of these ideas to atoms composed of many more electrons must be provided.

The results of a method known as the central field approximation verify their extension to atoms with more complex electron configurations. In this method, the potential function sums both the attractions between the nucleus and all of the electrons, neglecting any nuclear effects, and the individual electron-electron potentials, neglecting magnetic effects arising from electron motion. As in the case of the perturbation method, exact solutions to Schrödinger's equation can be obtained only for spherical potential fields. This approximation can be made since energy levels which are completely filled, assuming the Pauli exclusion principle, are very nearly spherical. This provides the basis for considering the electrons to be in a symmetrical potential in a way analogous to that of the hydrogen atom. Electrons closer to the nucleus are acted upon by a much greater potential than outer electrons since the latter are shielded from the nucleus, and thus are subject to a smaller attractive potential. A given electron is thus considered as being influenced by a net nuclear charge. In this way, the model approximates a symmetry of potential like that of a hydrogen atom. This simplification gives a potential function which includes the resultant of the nuclear and screening effects for each electron. An expression for each electron in the atom is contained in the Hamiltonian. Each of these equations is similar to that for hydrogen, so hydrogen-like wave functions can be used. The ionization energies of electrons calculated in this way are approximate. However, these are in sufficient agreement with the experiment to ensure that the hydrogen-like wave functions and their corresponding quantum numbers are applicable to atoms with many electrons. Therefore, the quantum numbers characterized earlier for the hydrogen atom may be applied to atoms with complex electron configurations in conjunction with the Pauli exclusion principle.

3.11.1. Electron Energy States

The permissible values for the four quantum numbers must be used in accordance with the Pauli exclusion principle. This acts as a guide to the way in which the electron energy states are filled, as shown in Table 3-2. Consider, for example, a given set of 2p electrons. Here $n = 2$ and $\ell = 1$. Then, m_ℓ can take the values -1, 0, 1 and $m_s = \pm 1/2$. From these factors it follows that only six states are possible:

State	2p	$2p^2$	$2p^3$	$2p^4$	$2p^5$	$2p^6$
m_s	$-1/2$	$-1/2$	$-1/2$	1/2	1/2	1/2
m_ℓ	-1	0	1	-1	0	1

Table 3-2
ALLOWED COMBINATIONS OF QUANTUM NUMBERS

n	ℓ	Designation	m_ℓ	m_s	Number	Total
1	0	1s	0	±1/2	2	2
2	0	2s	0	±1/2	2	8
2	1	2p	−1,0,1	±1/2	6	
3	0	3s	0	±1/2	2	
3	1	3p	−1,0,1	±1/2	6	18
3	2	3d	−2,−1,0,1,2	±1/2	10	
4	0	4s	0	±1/2	2	
4	1	4p	−1,0,1	±1/2	6	
4	2	4d	−2,−1,0,1,2	±1/2	10	32
4	3	4f	−3,−2,0,1,2,3	±1/2	14	

Note: Up to n = 4.

This is in agreement with the prior statement that the maximum number of states for any value of ℓ is given by $2(2\ell + 1)$. In a similar way, the s level ($\ell = 0$) can accommodate two electrons, the d level ($\ell = 2$) can hold 10 states, and the f level ($\ell = 3$) can accept at most 14 electrons.

Thus, for a given n and ℓ, only specified numbers of electrons can be accommodated. If this were not the case, both multiple occupancy of a given state and the case in which all electrons might be at one lowest level might exist (see sections on Statistics and Exclusion in Chapter 5). Under such conditions the observed spectral behavior and other phenomena noted in Section 3.9 could not have been explained.

The application of the Pauli exclusion principle to electrons as they are successively added to nuclei leads to the organization of atoms in the same way as they are given in the Periodic Table.

3.11.2. Ground State of Atoms

The electrons are most tightly bound to the nucleus in the ground state (state of lowest total energy) of isolated atoms. Therefore, this condition represents the most stable electron configuration.

In terms of the Pauli exclusion principle, the four quantum numbers of the electron states must be different. As additional electrons are added, the states available for occupation may be filled only according to the rules governing the quantum numbers. As the atomic number increases, n increases and the "radius" of the "orbit" increases, while the energy binding an electron to the nucleus decreases. In other words, as n increases, the electron quantum numbers change accordingly and the probability of finding an electron near the nucleus decreases. However, as the atomic number increases, the probability pattern shrinks somewhat because of the greater attractive force of the nucleus. The wave functions and their resultant probability densities play an important part in the determination of the electron configurations of atoms.

The hydrogen atom has one electron. This can occupy either of the states $n = 1$, $\ell = 0$, $m_\ell = 0$ and $m_s = \pm 1/2$. These are 1s states since $\ell = 0$. The energy difference between these states is small. Either one of these states, but not both, may be occupied in a given atom.

Helium, with an atomic number (Z) of two, has both 1s states occupied. This is a very stable configuration and accounts for its noble behavior. The greater charge on the nucleus causes the maximum probability of finding an electron to be closer to the nucleus than in the case of hydrogen.

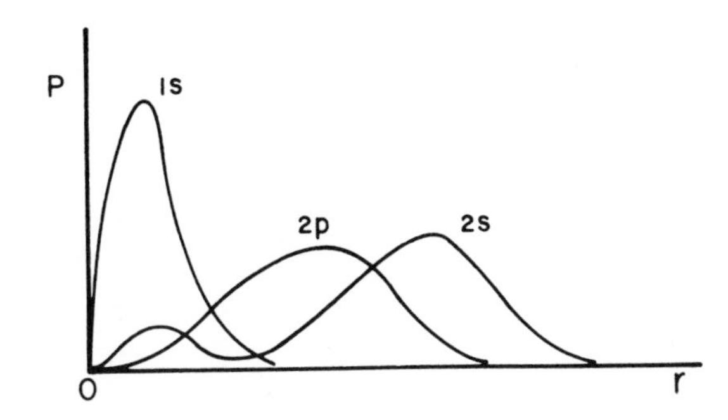

FIGURE 3-12. Relative probability densities for a lithium atom.

Lithium, Z = 3, has two electrons in $1s^2$ states as does He, and possesses one outer electron. This will occupy a 2s state rather than a 2p state. This arises from the fact that the 2s wave function can penetrate to the nucleus and the 2p does not. It will have a lower angular momentum and, therefore, a lower energy. The 2s state, accordingly, will be more stable. This is shown in Figure 3-12.

Beryllium has a configuration like that of Li plus an additional outer electron. The additional electron fills the 2s level to give it a $1s^2\ 2s^2$ configuration. This does not mean that Be should behave nobly as does He because its 2s states are completely filled. The overlap of the 2s and 2p probability functions permits the 2s electrons to enter into chemical and physical processes (see Chapter 5).

Boron, Z = 5, has a Be core with an additional electron. The fifth electron might occupy either a 2p or 3s state. However, while the 3s wave function represents greater penetration to the core, it implies a greater distance from the core, and thus, a higher energy configuration. For this reason, the 2p state is the most likely one to be occupied and the configuration of B is given by $1s^2\ 2s^2\ 2p^1$ (see Table 3-3).

Carbon, Z = 6, follows the behavior of B and has a $1s^2\ 2s^2\ 2p^2$ configuration. As in the case of Be, the overlap of the 2s and 2p probability densities accounts for much of its behavior. The polymorphic forms of C and its chemical and physical properties and bonding can be explained in terms of this s-p overlap. This overlap condition is also known as s-p hybridization.

The elements nitrogen, oxygen, and fluorine each successively acquire an additional 2p electron to the configuration of C to give $1s^2\ 2s^2\ 2p^3$, $1s^2\ 2s^2\ 2p^4$, and $1s^2\ 2s^2\ 2p^5$ configurations, respectively.

Neon with 10 electrons, has a completed 2p level and forms a $1s^2\ 2s^2\ 2p^6$ array. The p level is completed (n = 2, ℓ = 1) and the number of combinations is $2(2\ell + 1) = 6$. Here, the electrons of the completed p level are very tightly bound to the nucleus and noble-gas behavior, like that of He, results. This highly stable electron array resists interactions with other elements so that it does not normally form compounds. This inert behavior of Ne, and gases with similar p^6 configurations, is the reason for the name ''noble gases'' given to these elements.

The elements Na through Ar continue to fill up the 3s and 3p states in a regular way, similar to that previously described, starting with $1s^2\ 2s^2\ 2p^6\ 3s^1$ for Na and ending with $1s^2\ 2s^2\ 2p^6\ 3s^2\ 3p^6$ for Ar, where argon has the noble-gas configuration.

Potassium, Z = 19, might be expected to have its last electron either in the 3d or 4s state. But, as previously noted, the 4s wave function has a better probability of core penetration than the 3d, but the 3d represents greater orbital momentum. The 4s state therefore is a lower energy configuration than the 3d. Hence, the 4s state is occupied in preference to the 3d. The same is true for Ca.

Table 3-3
GROUND STATE ELECTRON CONFIGURATIONS OF ATOMS

Z[a]	Element	Outer configuration	Z[a]	Element	Outer configuration
1	H	1s	50	Sn	$5s^2\ 5p^2$
2	He	$1s^2$	51	Sb	$5s^2\ 5p^3$
			52	Te	$5s^2\ 5p^4$
3	Li	2s	53	I	$5s^2\ 5p^5$
4	Be	$2s^2$	54	Xe	$5s^2\ 5p^6$
5	B	$2s^2\ 2p$			
6	C	$2s^2\ 2p^2$	55	Cs	6s
7	N	$2s^2\ 2p^3$	56	Ba	$6s^2$
8	O	$2s^2\ 2p^4$	57	La	$5d\ 6s^2$
9	F	$2s^2\ 2p^5$	58	Ce	$4f^2\ 6s^2$
10	Ne	$2s^2\ 2p^6$	59	Pr	$4f^3\ 6s^2$
			60	Nd	$4f^4\ 6s^2$
11	Na	3s	61	Pm	$4f^5\ 6s^2$
12	Mg	$3s^2$	62	Sm	$4f^6\ 6s^2$
13	Al	$3s^2\ 3p$	63	Eu	$4f^7\ 6s^2$
14	Si	$3s^2\ 3p^2$	64	Gd	$4f^7\ 5d\ 6s^2$
15	P	$3s^2\ 3p^3$	65	Tb	$4f^8\ 5d\ 6s^2$
16	S	$3s^2\ 3p^4$	66	Dy	$4f^9\ 5d\ 6s^2$
17	Cl	$3s^2\ 3p^5$	67	Ho	$4f^{10}\ 5d\ 6s^2$
18	Ar	$3s^2\ 3p^6$	68	Er	$4f^{11}\ 5d\ 6s^2$
			69	Tm	$4f^{12}\ 5d\ 6s^2$
19	K	4s	70	Yb	$4f^{13}\ 5d\ 6s^2$
20	Ca	$4s^2$	71	Lu	$4f^{14}\ 5d\ 6s^2$
21	Sc	$3d\ 4s^2$	72	Hf	$5d^2\ 6s^2$
22	Ti	$3d^2\ 4s^2$	73	Ta	$5d^3\ 6s^2$
23	V	$3d^3\ 4s^2$	74	W	$5d^4\ 6s^2$
24	Cr	$3d^5\ 4s$	75	Re	$5d^5\ 6s^2$
25	Mn	$3d^5\ 4s^2$	76	Os	$5d^6\ 6s^2$
26	Fe	$3d^6\ 4s^2$	77	Ir	$5d^9$
27	Co	$3d^7\ 4s^2$	78	Pt	$5d^9\ 6s^1$
28	Ni	$3d^8\ 4s^2$			
			79	Au	$5d^{10}\ 6s$
29	Cu	$3d^{10}\ 4s$	80	Hg	$6s^2$
30	Zn	$4s^2$	81	Tl	$6s^2\ 6p$
31	Ga	$4s^2\ 4p$	82	Pb	$6s^2\ 6p^2$
32	Ge	$4s^2\ 4p^2$	83	Bi	$6s^2\ 6p^3$
33	As	$4s^2\ 4p^3$	84	Po	$6s^2\ 6p^4$
34	Se	$4s^2\ 4p^4$	85	At	$6s^2\ 6p^5$
35	Br	$4s^2\ 4p^5$	86	Rn	$6s^2\ 6p^6$
36	Kr	$4s^2\ 4p^6$			
			87	Fr	7s
37	Rb	5s	88	Ra	$7s^2$
38	Sr	$5s^2$	89	Ac	$6d\ 7s^2$
39	Y	$4d\ 5s^2$	90	Th	$6d^2\ 7s^2$
40	Zr	$4d^2\ 5s^2$	91	Pa	$6d^3\ 7s^2$
41	Nb	$4d^4\ 5s$	92	U	$6d^4\ 7s^2$
42	Mo	$4d^5\ 5s$	93	Np	$5f^6\ 7s^2$
43	Tc	$4d^6\ 5s$	94	Pu	$5f^5\ 6d\ 7s^2$
44	Ru	$4d^7\ 5s$	95	Am	$5f^6\ 6d\ 7s^2$
45	Rh	$4d^8\ 5s$	96	Cm	$5f^7\ 6d\ 7s^2$
46	Pd	$4d^{10}$	97	Bk	$5f^8\ 6d\ 7s^2$
			98	Cf	$5f^9\ 6d\ 7s^2$
47	Ag	$4d^{10}\ 5s$	99	Es	$5f^{10}\ 6d\ 7s^2$
48	Cd	$5s^2$	100	Fm	$5f^{11}\ 6d\ 7s^2$
49	In	$5s^2\ 5p$			

[a] Atomic number.

Beyond Ca, with the exceptions of Cr and Cu, the 4s states are completely filled and the successive electrons fill 3d states. These are the next lowest energy states which are available. The elements Sc through Ni constitute the elements of the first transition period, since the 3d levels fill in a regular way, with the exception of Cr, after the 4s states are filled. At Cu the d band is filled with 10 electrons and the configuration is $1s^2\ 2s^2\ 2p^6\ 3s^2\ 3p^6\ 3d^{10}\ 4s^1$. The outer configuration of Cr is $3d^5\ 4s^1$ which gives a stable half-filled d band. Had this been a $3d^6$ configuration, the required spin reversal in the remaining d half-band would have resulted in a higher energy, less stable array of electrons.

From Cu to Kr the 4s and 4p electrons are added in the regular way. At Kr the 4p levels are completely filled and form the noble-gas array.

The elements from Kr to Pd fill 4d and 5s states as additional electrons are added in a way analogous to the first transition series. Here the d states fill uniformly, but the s states contain only one electron from Nb on, except for Y ($4d\ 5s^2$) and Zr ($4d^2\ 5s^2$). At Pd the d states are completed at 10 and no s states are occupied in the isolated atom.

Atoms from Ag to Xe have occupied 5s and 5p electron states in which the p levels increase in a regular way and end with the noble-gas configuration of $5p^6$ at Xe. The elements Cs and Ba have added 6s electrons in addition to those of the Xe core. With the exception of La, the 4f states of atoms from Ce to Eu acquire electrons and from Gd to Yb, where a single 5d state is present in each atom, the 4f states continue to fill. In the case of La a 5d state is occupied, but the 4f state of La is vacant. The series from Lu through Pt represents another transition series in which the 4f states are filled and the 5d states fill progressively, and all but Pt and Ir have $6s^2$ levels. At Au all of the inner states are completed and the configuration includes one outer 6s state.

From Hg through Rn the 6p states are successively filled until at Rn the noble-gas configuration of $6p^6$ occurs. The 7s states fill at Fr (7s) and Ra ($7s^2$), and Ac, Th, Pa and U ($6d^4\ 7s^2$) add successive 6d electrons to the $7s^2$ states. Np and Pu have $5f^5\ 7s^2$ and $5f^5\ 6d\ 7s^2$ arrays. Am through Fm add successively increasing numbers of electrons in the 5f states while retaining $6d\ 7s^2$ outer configurations.

The valence electrons in outer, incompletely filled shells are those which normally enter into chemical reactions and determine the types of compounds thus formed. Many such compounds have stoichiometries which are governed by the valences as given for the ground state.

For most purposes of this text, the inner electrons occupying completed shells, are considered to be tightly bound to the nucleus; these do not ordinarily enter into physical processes. However, these bound electrons must be taken into consideration in order to explain diamagnetic properties (Section 8.2.1 in Chapter 8). The valence electrons are much more loosely bound; consequently, these participate in many physical processes. Therefore, the valence electrons are of major interest in considering the physical properties of solids.

Normal elements (those with completed inner shells and partly filled outer levels) in the crystalline state frequently do not show the integral valences which would be expected from their ground-state configurations. Normal metallic bonding is described in Section 10.6 in Chapter 10. Another of the reasons for nonintegral valences arises from the concept that some of these electrons may be involved primarily in the bonding of the atoms of the crystal; when this is the case, they are not normally available for engaging in physical processes. When this behavior is considered, the number of available, or uninvolved, electrons per atom also results in a nonintegral valence for the atom. An additional reason for nonintegral valences results from the fact that some of the outer electrons can resonate to other unfilled levels in the atoms composing the solids. This is particularly true of transition elements and alloys involving transition elements. Nickel provides a good example of this. In the ground state elemental Ni has a $3d^8 \ 4s^2$ array. In the crystalline state the configuration of Ni is considered to be $3d^{9.4} \ 4s^{0.6}$ as a result of this effect. In the case of alloys of normal elements with transition elements, the valence electrons from the normal elements tend to occupy d states of the transition element. Strong evidence for this is shown by the thermoelectric and magnetic properties of these alloys. More detailed descriptions of this behavior are provided in Sections 7.9.3 and 7.9.4 in Chapter 7, and Sections 9.9 and 9.10 in Chapter 9.

3.12. PROBLEMS

1. Show the equivalence of Equations 3-3 and 3-4 to Equations 3-10 and 3-12.
2. Compare the conditions given for meaningful solutions to Schrödinger's equation to the conditions necessary for a fair "shell" game.
3. Obtain a graphical solution for the equation given for an electron in a finite well (Section 3.7).
4. Can a crude analogy be made between ionic, covalent, and metallic type bonding and Figure 3-3?
5. Show that Equations 3-69 and 3-71 are solutions to Equations 3-68 and 3-70, respectively.
6. Make a first approximation of the energy changes which result from electron transitions from n = 2, 3, and 4 to n = 1.
7. Use the perturbation method to write the Hamiltonian for the helium atom neglecting the electron interaction term.
8. Use the perturbation method to write the Hamiltonian for the helium atom including the perturbation terms.
9. Describe how the central field approximation may be applied to the lithium atom.
10. Show, in tabular form, that the maximum numbers of d and f states cannot exceed 10 and 14, respectively.
11. Make a chart for the first 18 elements based upon their 4 quantum numbers.

3.13. REFERENCES

1. **Sproull, R. P.**, *Modern Physics,* John Wiley & Sons, New York, 1956.
2. **Richtmyer, F . K., Kennard, E. H., and Lauritsen, T.**, *Introduction to Modern Physics,* 5th ed., McGraw-Hill, New York, 1955.
3. **Stringer, J.**, *An Introduction to the Electron Theory of Solids,* Pergamon Press, Elmsford, N. Y., 1967.
4. **Hume-Rothery, W.**, *Atomic Theory for Students of Metallurgy,* The Institute of Metals, London, 1952.
5. **Gillespie, D. T.**, *A Quantum Mechanics Primer,* International Textbook, Scranton, Pa., 1970.
6. **Dekker, A. J.**, *Solid State Physics,* Prentice-Hall, Englewood Cliffs, N. J., 1959.
7. **Kittel, C.**, *Introduction to Solid State Physics,* John Wiley & Sons, New York, 1966.
8. **Hertzberg, G.**, *Atomic Spectra and Atomic Structure,* Dover, New York, 1944.
9. **Mahan, B. H.**, *College Chemistry,* Addison-Wesley, Reading, Mass., 1969.
10. **Raimes, S.**, *The Wave Mechanics of Electrons in Metals,* North-Holland, Amsterdam, 1961.
11. **Martin, T. C., Jr. and Leonard, W. F.**, *Electrons and Crystals,* Brooks/Cole, Monterey, Calif., 1970.

Chapter 4

THERMAL PROPERTIES OF NONCONDUCTORS

Atomic structure having been considered, the thermal properties of solids are introduced at this point because the thermal vibrations of the ions comprising a crystal affect other physical properties, including their electrical characteristics. As a means of simplifying the treatment of lattice vibrations, only the ions of pure, elemental solids will be considered. The additional contributions of the electrons to thermal and other physical properties are discussed in subsequent chapters.

The simplest way of treating ions in this manner is to consider the atoms in the crystal to be covalently bonded. Under this condition the valence electrons are bonded very strongly to the atoms, forming ions. Crystals bonded in this way show minimal electrical and thermal conductivities; they are classed as insulators because of the extremely low number and mobility of any valence electrons which might be available. In other words, these materials are essentially nonconductors. Intrinsic semiconductors also may be included for those temperature ranges in which the electron population of the conduction band (Chapter 11, Volume III) is negligibly small.

The thermally induced vibrations of the ions are responsible for such physical properties as heat capacity, temperature coefficient of thermal expansion, and thermal conductivity. All of these are important physical properties of materials.

While not included in this text, the understanding of thermally induced lattice vibrations provides a basis for a better comprehension of many solid-state reactions which include diffusion, allotropic changes, and order-disorder.

4.1. HEAT CAPACITY

On the basis of the foregoing introduction, this topic will consider only elemental solids. The heat capacity of a substance at constant volume is defined as

$$C_V = \left[\frac{\partial U}{\partial T}\right]_V \tag{4-1}$$

where U is the internal energy and T is the absolute temperature. The heat capacity at constant pressure is

$$C_P = \left[\frac{\partial H}{\partial T}\right]_P = \left[\frac{\partial (U + PV)}{\partial T}\right]_P \tag{4-2}$$

in which H is the enthalpy, P is the pressure, and V is the volume. Performing the indicated differentiation,

$$C_P = \left[\frac{\partial U}{\partial T}\right]_P + P\left[\frac{\partial V}{\partial T}\right]_P + V\left[\frac{\partial P}{\partial T}\right]_P$$

and noting that $\partial P/\partial T = 0$ at constant pressure,

$$C_P = \left[\frac{\partial U}{\partial T}\right]_P + P\left[\frac{\partial V}{\partial T}\right]_P \tag{4-3}$$

The quantity ∂U can be expressed as

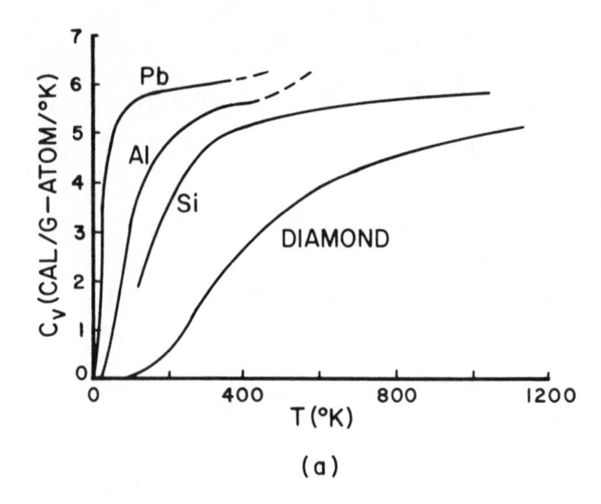

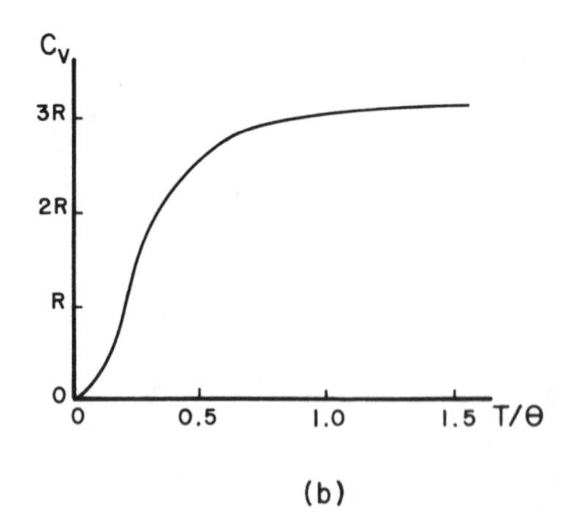

FIGURE 4-1. (a) Heat capacities of some elements. (After Richtmyer, F. K., Kennard, E. H., and Lauritsen, T., *Introduction to Modern Physics,* 5th ed., McGraw-Hill, New York, 1955, 410. With permission.) (b) The "universal" curve. (After Seitz, F., *Modern Theory of Solids,* McGraw-Hill, New York, 1940, 109. With permission.)

$$\partial U = \left[\frac{\partial U}{\partial V}\right]_T \partial V + \left[\frac{\partial U}{\partial T}\right]_V \partial T$$

and when divided by ∂T this becomes

$$\frac{\partial U}{\partial T} = \left[\frac{\partial U}{\partial V}\right]_T \left[\frac{\partial V}{\partial T}\right]_P + \left[\frac{\partial U}{\partial T}\right]_V \tag{4-4}$$

Thus, C_P (Equation 4-3) becomes

$$C_P = \left[\frac{\partial U}{\partial V}\right]_T \left[\frac{\partial V}{\partial T}\right]_P + \left[\frac{\partial U}{\partial T}\right]_V + P\left[\frac{\partial V}{\partial T}\right]_P \tag{4-5}$$

When this is used with Equation 4-1

$$C_P - C_V = \left[\frac{\partial U}{\partial V}\right]_T \left[\frac{\partial V}{\partial T}\right]_P + \left[\frac{\partial U}{\partial T}\right]_V$$

$$+ P\left[\frac{\partial V}{\partial T}\right]_P - \left[\frac{\partial U}{\partial T}\right]_V$$

$$C_P - C_V = \left[\frac{\partial U}{\partial V}\right]_T \left[\frac{\partial V}{\partial T}\right]_P + P\left[\frac{\partial V}{\partial T}\right]_P$$

$$= \left\{\left[\frac{\partial U}{\partial T}\right]_T + P\right\}\left[\frac{\partial V}{\partial T}\right]_P \tag{4-6}$$

For most solids, especially metals at normal pressures, $(\partial V/\partial T)_P$ is relatively small, $\Delta V/\Delta T$ being of the order of 10^{-4} to 10^{-5} cm^3/°C. So, for many purposes it can be assumed that this quantity is negligible and that

$$C_P \simeq C_V \tag{4-7}$$

This approximation will be employed for solids only where applicable and convenient. It gives a maximum error of less than 0.5 cal/mol K.

Dulong and Petit (1818) showed that the specific heats of many substances were related to their atomic weights; the product of their specific heats and atomic weights being approximately constant, about 6 cal/mol deg. This is only an approximation. For many metals, the heat capacity lies between 5 and 7 cal/mol deg at 0°C, and has an average value of 6.2. The heat capacities of metals are not constant but increase about 0.04%/°C for temperatures above 0°C. This is shown in Figure 4-1. The approximate nature of the Dulong and Petit "law" is apparent from the figure. Nevertheless, it is a helpful rule-of-thumb in practical applications. It is also apparent from the figure that the curves for the elements shown all have the same type of sigmoidal-shaped curve.

The curves of each of the solids shown in Figure 4-1 can be made to fall upon a single, "universal" curve. This is done by means of a parameter which is specific for each element, the characteristic temperature, θ, which is used to normalize the curves. In the very low temperature range, approximately to $\theta/10$ for many materials, C_V varies as T^3. When the data for C_V are plotted versus T/θ, the data fall very close to a common curve. It should be noted that many exceptions to this curve exist. It is really only valid for elemental, nonmolecular solids. It is not valid where allotropic phase or magnetic changes take place. It is also sensitive to the presence of order-disorder reactions in alloys. Peaks and discontinuities in heat capacity curves have been employed to detect such changes and reactions in the solid state.

The high-temperature portion of the "universal" curve and the rule of Dulong and Petit can be explained by showing that the average energy of an atom is approximately constant at elevated temperatures. This is done by dividing the average energy of a mole of a solid by Avogadro's number: 6 cal/mol K $\div$ 6.02×10^{23} atoms/mol $\simeq 1 \times 10^{-23}$ cal/atom K, an approximation in good agreement with the experimental results. As is shown in subsequent treatments, the low-temperature behavior requires more complex explanations.

It should be noted that some elements with low atomic weights have heat capacities less than the predicted values. Among these are: Li 5.0, Be 5.2, B 4.6, C 3.2, and Li 5.0 cal/mol K. Others, such as Na and K have values of 6.7 and 7.1 cal/mol K, respectively.

Kopp's rule may be used to approximate the heat capacity of a solid as the sum of the heat capacities of its constituent atoms. The accuracy of this estimation improves with increasing temperature.

4.1.1. Classical Theory

The law of Dulong and Petit can be explained by means of classical physics. The classical theory assumed that the internal energy of a solid could be considered to reside in its ion cores. A solid was thought of as being made up of an assembly of ion cores which behaved as simple harmonic oscillators, vibrating about an equilibrium position, in thermal equilibrium at a given temperature. This, of course, neglects any contribution of the valence electrons to the internal energy of the solid. The thermal equilibrium of the ion cores is treated, as in the case of ideal gases, as though the energy distribution is continuous and makes use of the Maxwell-Boltzmann equipartition of energy.

The ions were considered to have three degrees of freedom which correspond to their energies of translation parallel to the three cartesian coordinates. The average energy per degree of freedom is

$$U = \frac{1}{2}k_B T \tag{4-8}$$

where k_B is Boltzmann's constant. The vibrating ion in the solid will possess both kinetic energy and potential energy (a restoring energy) for each degree of freedom. The averages of these energies will be equal. Thus the total energy for each degree of freedom will be k_BT. So, for the three degrees of freedom the average energy of each ion will be 3 k_BT. For a mole, the internal energy will be

$$U = 3\,Nk_B T \tag{4-9}$$

where N is Avogadro's number (6.02×10^{23} atoms/mol). From Equation 4-1 the heat capacity should be

$$C_V = \left[\frac{\partial U}{\partial T}\right]_V = 3\,Nk_B \tag{4-10}$$

It will be recalled that $Nk_B = R$, the gas constant (1.9872 cal/mol degree). This provides a value of C_V for a metal equal to approximately 6 cal/mol degree, a value which agrees with the findings of Dulong and Petit. However, this fails to provide any temperature dependence of the heat capacity.

Implicit in this oversimplified treatment is the idea that each ion in the solid is treated as an independent oscillator. It is apparent that the oscillations of a given ion will affect those of its neighbors. Models of heat capacity which are more representative of actual conditions must account for such ion-ion interactions.

4.1.2. The Einstein Model

Einstein (1906) developed a model which overcame the failure of the classical approach to describe the heat capacity of a solid as a function of temperature. Instead of using Equation 4-8, he used the then new concept of expressing energy in terms of the frequency of the oscillating ion (Equation 1-6). By means of the Planck idea of discrete energies, the average energy of one such oscillator is found by obtaining the average from the energies of all of the ions in the solid. This average energy is given by

$$\bar{E} = \frac{\sum_{o}^{\infty} nh\nu \exp(-nh\nu/k_BT)}{\sum_{o}^{\infty} \exp(-nh\nu/k_BT)} \tag{4-11}$$

Rewriting the sums

$$\bar{E} = \frac{h\nu\left(0 + e^{-\frac{h\nu}{k_BT}} + 2e^{-2\frac{h\nu}{k_BT}} + + +\right)}{\left(1 + e^{-\frac{h\nu}{k_BT}} + e^{-2\frac{h\nu}{k_BT}} + + +\right)} \tag{4-12}$$

For convenience, let $x = -h\nu/k_BT$. Then Equation 4-12 becomes

$$\bar{E} = \frac{h\nu\,(o + e^x + 2e^{2x} + + +)}{(1 + e^x + e^{2x} + + +)} \tag{4-13}$$

It will be seen that the numerator is the derivative of the denominator of Equation 4-13, so it can be expressed as

$$\bar{E} = h\nu \frac{d}{dx} \ln(1 + e^x + e^{2x} + + +) \tag{4-14}$$

The series within the parentheses is given by

$$1 + e^x + e^{2x} + + + = \frac{1}{1 - e^x} \tag{4-15}$$

So, the average energy of one such oscillator is given by

$$\bar{E} = h\nu \frac{d\ln}{dx}\left[\frac{1}{1 - e^x}\right] \tag{4-16}$$

When the derivative is taken

$$\bar{E} = h\nu \frac{e^x}{1 - e^x} \tag{4-17}$$

This can be further expressed as

$$\bar{E} = h\nu \frac{1}{e^{-x} - 1} \tag{4-18}$$

or, when the value of the exponent is substituted

$$\bar{E} = \frac{h\nu}{e^{h\nu/k_BT} - 1} \tag{4-19}$$

An expression is thus obtained for the average energy of an oscillator; this is quite different from that given by Equation 1-6.

It will be noted that Equation 4-19 equals the classical case at high temperatures. This may be shown by expressing the exponential term as the first two terms of a series. Thus

$$\frac{h\nu}{e^{h\nu/k_BT} - 1} \simeq \frac{h\nu}{1 + \frac{h\nu}{k_BT} - 1} = h\nu \cdot \frac{k_BT}{h\nu} = k_BT \tag{4-20}$$

In order to apply this average energy, Einstein based his ideas upon the classical picture of a solid and extended these ideas. As in the classical case, it was assumed that the internal energy of a solid was associated with the ions only. The energy of the electrons was not taken into account. The solid thus was treated as an assembly of *independent* simple-harmonic oscillators in thermal equilibrium. That is, each ion of the solid independently oscillated with the same given frequency, ν_o. As noted for the classical case, this is too great an oversimplification because each oscillating ion at least would be expected to affect the oscillations of its nearest neighbors so that their frequencies could not possibly be all the same. The inclusion of the quantum-mechanic approach, represented by Equation 4-19, did constitute major progress.

If each ion of the solid has three translational degrees of freedom and N ions constitute a mole, then, by Equation 4-19, the internal energy U of a mole of the solid is given by the sum of the average energies of all of the ions in the solid:

$$U = \frac{3N\, h\nu_o}{\exp(h\nu_o/k_BT) - 1} \tag{4-21}$$

neglecting zero point energy, and differentiating

$$C_v = \left.\frac{\partial U}{\partial T}\right|_V = \frac{3N\,(h\nu_o)\,(\exp h\nu_o/k_BT)\,(h\nu_o/k_BT^2)}{\left[\exp(h\nu_o/k_BT) - 1\right]^2} \tag{4-22}$$

The numerator is multiplied and divided by k_B to obtain

$$C_v = \frac{3Nk_B \exp(h\nu_o/k_BT)\,(h\nu_o/k_BT)^2}{\left[\exp(h\nu_o/k_BT) - 1\right]^2} \tag{4-23}$$

Expanding the exponential terms as the first two terms of a series

$$C_v \approx \frac{3Nk_B\,(1 + h\nu_o/k_BT)\,(h\nu_o/k_BT)^2}{\left[1 + h\nu_o/k_BT - 1\right]^2} = 3Nk_B\left(1 + \frac{h\nu_o}{k_BT}\right) \tag{4-24}$$

As the temperature increases, C_v approaches the $3Nk_B$ limit, the classical value. Thus, the Einstein model agrees with the findings of Dulong and Petit at high temperatures.

How does Einstein's model behave at lower temperatures? Starting with Equation 4-23, and expanding the denominator

$$C_v = \frac{3Nk_B \exp(h\nu_o/k_BT)\,(h\nu_o/k_BT)^2}{\exp(2h\nu_o/k_BT) - 2\exp(h\nu_o/k_BT) + 1}$$

and dividing both numerator and denominator by $\exp(h\nu_o/k_BT)$

$$C_v = \frac{3Nk_B(h\nu_o/k_BT)^2}{\exp(h\nu_o/k_BT) - 2 + \exp(-h\nu/k_BT)} \tag{4-25}$$

As the temperature approaches zero, the negative exponential term in the denominator

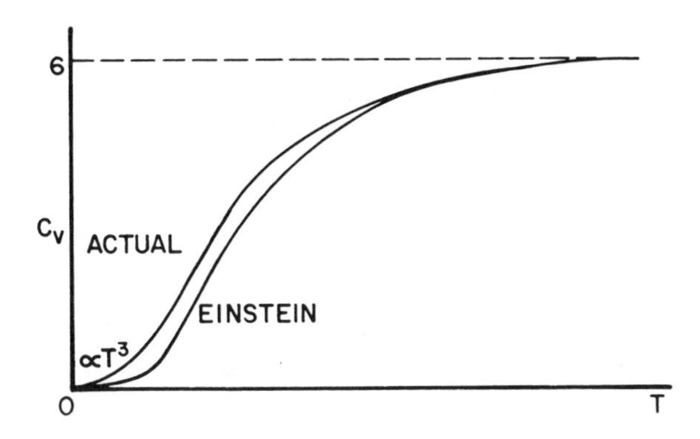

FIGURE 4-2. Comparison of the Einstein results for heat capacity with experimental behavior.

will approach zero, while the positive one becomes very large. This will become large at a much faster rate than the fraction in the numerator. This will cause C_v to approach zero in an exponential fashion. Thus the Einstein model fits the observed conditions in the limits. Figure 4-2 shows how this fits the experimental data over a range of temperatures. It is apparent that the Einstein relationship gives a poor fit in the lower range of temperatures because of its exponential rather than T^3 behavior; this improves as the temperature increases. This degree of agreement is surprisingly good considering the simplicity of the model.

The major difficulty with this model is that it was assumed that all of the ions oscillated with one given frequency ν_o. It is emphasized again that this condition cannot exist because each oscillating ion and its neighbors mutually interact, and so on, throughout the solid. Thus, each ion of the lattice affects and is affected by every other ion of the lattice.

It should be noted, however, that the Einstein model is useful for approximating lattice vibrational effects. This is done at temperatures above the range in which the exponential effects are pronounced. Such approximations usually are made close to, or above, the Debye temperature (see Section 4.1.4 and Table 4.1, Chapter 4) to simplify the calculations. An illustration of this usage is given in the analysis of Matthiessen's rule (Section 6.2, Chapter 6).

This model continues to have other important applications. One of these is to calculate the heat capacities of gases when information is incomplete or unavailable. Ideal behavior is assumed and the internal energy is considered to consist of three components. Two of these, the translational and rotational modes are specified by 1/2RT per mole for each degree of freedom. The vibrational modes are included by appropriate summations of Equation 4-19 over all applicable modes. The heat capacity is determined by the derivative of the sum of the operative components with respect to temperature. Tables are available for evaluating the Einstein summations. This treatment gives reasonable approximations at, or above, room temperature.

4.1.3. The Born-von Kármán Model

The Born-von Kármán (1912) model represents an important advance in the attempt to overcome the major failing of the Einstein approach; a means for accounting for ion-ion interactions is included. Like the other models, it assumed that the internal energy of a solid resided in the ions and no account was made of the contributions of the electrons. In this case, however, the solid was treated as an assembly of *coupled,* simple-harmonic oscillators. This concept represented an important advance. And, as

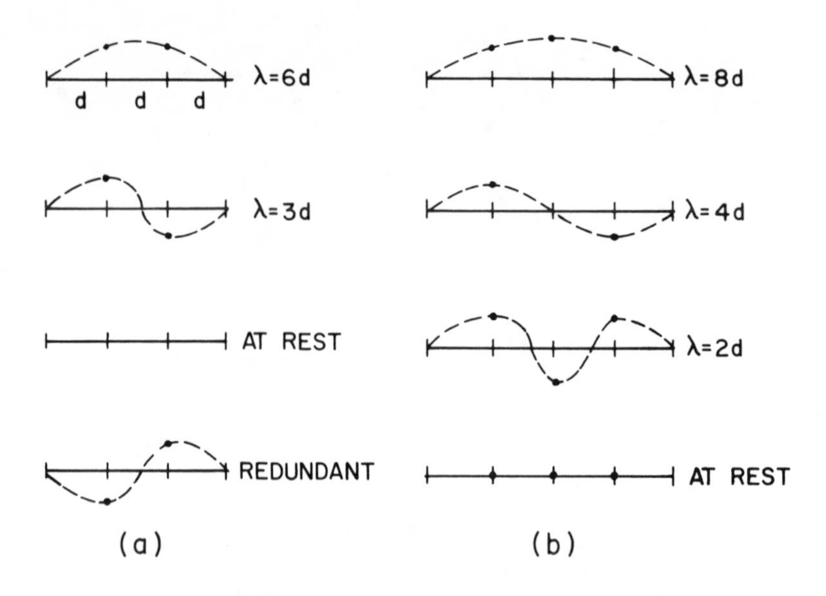

FIGURE 4-3. Sketches of vibrating particles on strings.

in Einstein's model, Equation 4-19 was used for the average energy of the ion. But, the big difference here is that the frequency, ν_i, is that of each of the individual ions. Considering that each oscillating ion has three degrees of freedom and that the solid consists of N ions, the internal energy of the solid is determined by the sum of the energies of all of the oscillating ions:

$$U = \sum_{i=1}^{3N} \frac{h\nu_i}{\exp(h\nu_i/k_B T) - 1} + \text{zero point energy} \tag{4-26}$$

The task is now one of evaluating the factor ν_i and to describe the way in which the vibrations of the coupled oscillators vary.

4.1.3.1. Vibrational Modes

The coupled ions of the Born-von Kármán model can be pictured as particles on a string. Consider two such particles (Figure 4-3a).

In the first case, the two vibrating particles are in synchronization; their wavelength is 6d, where d is the interionic distance. The vibrations of the particles are exactly 180° out of phase in the second case and their wavelength is 3d. The third case shows the particles at rest. The fourth case again shows the particles 180° out of phase; this is redundant. Similar cases are shown in Figure 4-3b for three particles. This behavior may be extended to include any number of oscillating particles.

Vibrations of this kind are called normal modes of vibration and are like standing waves because the ends are fixed. It is possible to have any number of particles. Any complex ionic vibrations can be described by the addition, or superposition, of the independent vibrational motions of atoms in normal modes. These normal modes are preserved independently of other modes of motion. This permits their addition.

It is easy to see that, in the plane of the paper, one particle on the string would have one normal mode, one wavelength, and one frequency. Two ions would have two normal modes, two wavelengths, and two frequencies. Three ions would have three normal modes, three wavelengths, and three frequencies. N ions would have N normal modes, N wavelengths, and frequencies. The other two-vibrational planes must be con-

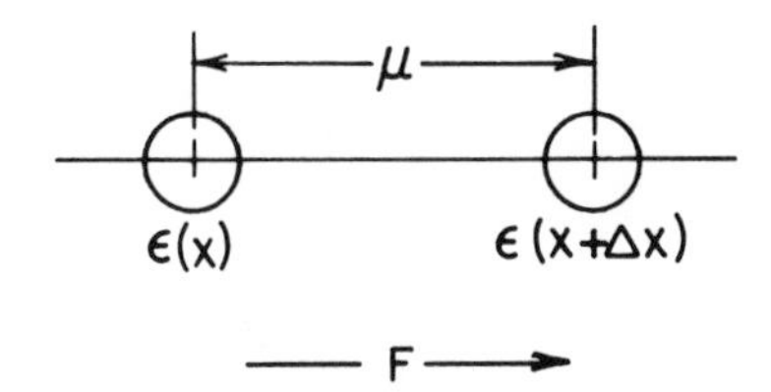

FIGURE 4-4. Displacement of an ion on a string.

sidered also. Thus, for N particles with oscillations polarized in three planes, 3N normal vibrational modes exist.

If ω, the angular frequency, is used instead of ν in Equation 4-19, where $\omega = 2\pi\nu$, then the average energy of an oscillating ion is given by

$$U = \frac{\hbar\omega}{\exp(\hbar\omega/k_BT) - 1} \tag{4-27}$$

where $\hbar = h/2\pi$. If the number of vibrational modes of the ions were known, the product of this quantity and the average energy of each of the ions would give the average internal energy of the solid. Assume, then, that in the frequency range between ω and $\omega + d\omega$ there are $N(\omega)d\omega$ vibrational modes. This is known as the density of vibrational modes. Now, if the number of oscillating ions is sufficiently large, it is possible to integrate over the range of angular frequencies instead of summing as in Equation 4-26. This would give the internal energy of the solid as

$$U = \int_{\omega_{min}}^{\omega_{max}} \frac{\hbar\omega}{\exp(\hbar\omega/k_BT) - 1} N(\omega)\, d\omega \tag{4-28}$$

This has simplified the problem, but it is now necessary to define the distribution function $N(\omega)$, or density of vibrating states. In order to do this it will be necessary to examine the oscillations of the coupled particles comprising the Born-von Kármán model of a solid.

4.1.3.2. The Vibrating String

The Born-von Kármán treatment thus involves integrating over all of the waves of strings of ions in the solid. In order to accomplish this, a start will be made by considering the simplest case: that of the vibrations in a *homogeneous,* one-dimensional line of ions of linear density, ϱ. Only longitudinal waves will be involved; i.e., the motion of each ion in the string will be parallel to the line itself. This behavior is shown in Figure 4-4. Let x be the coordinate of a given line element, and its displacement from its equilibrium position be μ. Then the strain, ε, or fractional change in length is $\partial\mu/\partial x$. If F is the force producing the strain, the elastic modulus, or stiffness, is given by

$$c = \frac{F}{\epsilon} \tag{4-29}$$

Now consider the forces on an element of the line, Δx. At one end of the element the strain is $\varepsilon(x)$, and is $\varepsilon(x + \Delta x)$ at the other end. The strain at $x + \Delta x$ is

$$\epsilon(x + \Delta x) = \epsilon(x) + \frac{\partial\epsilon}{\partial x}\Delta x \tag{4-30}$$

The strain was given by

$$\epsilon = \frac{\partial \mu}{\partial x} \qquad (4\text{-}31)$$

From this, by differentiation,

$$\frac{\partial \epsilon}{\partial x} = \frac{\partial^2 \mu}{\partial x^2} \qquad (4\text{-}32)$$

This is substituted into Equation 4-30 to give

$$\epsilon(x + \Delta x) = \epsilon(x) + \frac{\partial^2 \mu}{\partial x^2} \Delta x \qquad (4\text{-}33)$$

The net strain on the element is

$$\epsilon = \epsilon(x + \Delta x) - \epsilon(x) = \frac{\partial^2 \mu}{\partial x^2} \Delta x \qquad (4\text{-}34)$$

From Equation 4-29 the force acting on the line element is

$$F = \epsilon c = c \frac{\partial^2 \mu}{\partial x^2} \Delta x \qquad (4\text{-}35)$$

The force on the line element is also given by

$$F = ma = \rho \Delta x \frac{\partial^2 \mu}{\partial t^2} \qquad (4\text{-}36)$$

since its mass is linear and is given by $\varrho \Delta x$. Equations 4-35 and 4-36 are equated, and, since the length increment vanishes,

$$\frac{c}{\rho} \frac{\partial^2 \mu}{\partial x^2} = \frac{\partial^2 \mu}{\partial t^2} \qquad (4\text{-}37)$$

Let

$$\frac{c}{\rho} = A^2 \qquad (4\text{-}38)$$

Then

$$A^2 \frac{\partial^2 \mu}{\partial x^2} = \frac{\partial^2 \mu}{\partial t^2} \qquad (4\text{-}39)$$

Equation 4-39 is d'Alembert's equation (1747) for vibrating strings. It should be noted that

$$A = \left[\frac{c}{\rho}\right]^{1/2} = v_s \qquad (4\text{-}40)$$

which is the Newtonian expression for the velocity of sound, v_s (longitudinal waves) in a homogeneous medium.

A solution to Equation 4-39 may be found by trying

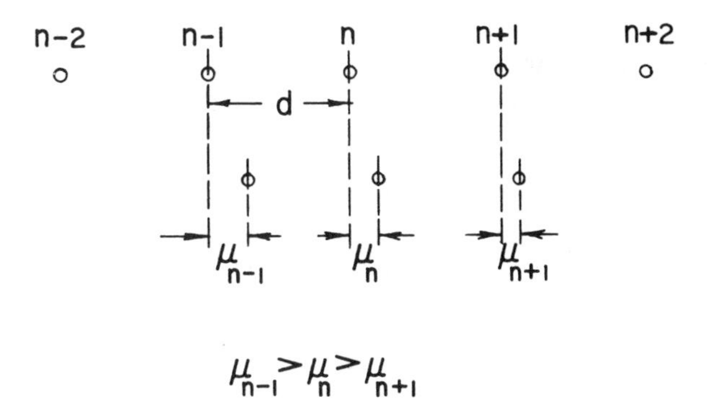

FIGURE 4-5. Equilibrium and displaced positions of a string of ions.

$$\mu = \exp\left[i(At + x)\right] \tag{4-41}$$

Equation 4-41 can be shown to be a solution of Equation 4-39 by differentiation and substitution.

The exponents of Equation 4-41 will be modified for future use. By Equation 4-40 and the substitutions noted below

$$A = v_s = \frac{\bar{k}}{\bar{k}} v_s = \frac{\bar{k}}{\bar{k}} \lambda\nu = \frac{\bar{k}}{\bar{k}} \frac{2\pi}{\bar{k}} \nu = \frac{\bar{k}\omega}{\bar{k}^2} = \frac{\omega}{\bar{k}}$$

and

$$x = \frac{\bar{k}x}{\bar{k}}$$

so that Equation 4-41 can be expressed as

$$\mu = D \exp\left[i\bar{k}(At + x)\right] \simeq D \exp\left[i(\omega t + \bar{k}x)\right] \tag{4-42}$$

where D is a constant. This will be the general form employed for the displacement induced by traveling longitudinal waves in the string. This expression will be applied later, as Equation 4-49, to derive the properties of waves in a one-dimensional array of ions.

4.1.3.3. Traveling Waves in One-Dimensional String

Some of the properties of the wave motion on a line of identical ions will now be derived. These will be more applicable to the Born-von Kármán model. In this case the string no longer is homogeneous. Here, the wave motion is similar to the elastic waves on a homogeneous string when the wavelength is much larger than the distance between the ions. An examination will also be made of the properties of waves of shorter wavelength.

The kind of displacement of interest is indicated in Figure 4-5. Only nearest neighbor interactions will be considered. The force acting on the nth ion is

$$F_n = \beta(\mu_{n+1} - \mu_n) - \beta(\mu_n - \mu_{n-1}) \tag{4-43}$$

in which β is the spring, or force, constant, μ_i are the displacements, and the quantities in the parentheses give the increases in the bond length between the designated ions. On a macroscopic scale, the line of ions has a linear density or

$$\rho = \frac{M}{d} \tag{4-44}$$

where the mass is given by M and d is the equilibrium distance between the ions. The force required to stretch a single bond between the ions is

$$F = \beta(\mu_n - \mu_{n-1}) = \beta \epsilon d \tag{4-45}$$

where ε is the unit strain. The modulus for this nonhomogeneous string is

$$\frac{F}{\epsilon} = \beta d \tag{4-46}$$

Equation 4-29 gave the modulus for the homogeneous string as

$$\frac{F}{\epsilon} = c$$

Equating these two relationships gives

$$\beta d = c \tag{4-47}$$

This equation will be useful in simplifying Equation 4-62 later.

Now, using Newton's second law, ma = F, equating Equations 4-36 and 4-43, and collecting the terms in Equation 4-43:

$$M \frac{d^2\mu}{dt^2} = \beta(\mu_{n+1} + \mu_{n-1} - 2\mu_n) \tag{4-48}$$

It is helpful to find a solution to Equation 4-48 similar to that obtained for traveling waves in the homogeneous string. However, since discrete particles are now involved, the quantity nd must be used instead of x in the exponent of Equation 4-42, and Equation 4-42 will take the form

$$\mu_n = \exp\left[i(\omega t + \bar{k}nd)\right] \tag{4-49}$$

It will be shown that a solution to Equation 4-48 is

$$-\omega^2 M = \beta(e^{i\bar{k}d} + e^{-i\bar{k}d} - 2) \tag{4-50}$$

This relationship will be used to define the properties of these waves. Upon differentiation of Equation 4-49

$$\frac{d\mu}{dt} = i\omega \exp\left[i(\omega t + \bar{k}nd)\right]$$

and

$$\frac{d^2\mu}{dt^2} = -\omega^2 \exp\left[i(\omega t + \bar{k}nd)\right] \qquad (4\text{-}51)$$

This expression is multiplied through by M to give an equation equivalent to Equation 4-48. Equating these two relationships gives

$$M\frac{d^2\mu}{dt^2} = -M\omega^2 \exp\left[i(\omega t + \bar{k}nd)\right]$$

$$= \beta\left(\mu_{n+1} + \mu_{n-1} - 2\mu_n\right) \qquad (4\text{-}52)$$

Now, Equation 4-49 may be substituted for the quantities within the parentheses on the right side of Equation 4-52 to obtain

$$\left(\mu_{n+1} + \mu_{n-1} - 2\mu_n\right) = e^{i\omega t}\left[e^{i\bar{k}(n+1)d} + e^{i\bar{k}(n-1)d} - 2e^{i\bar{k}nd}\right]$$

Now, replacing this expression in Equation 4-52

$$-M\omega^2 e^{i\omega t}\cdot e^{i\bar{k}nd} = e^{i\omega t}\left[e^{i\bar{k}(n+1)d} + e^{i\bar{k}(n-1)d} - 2e^{i\bar{k}nd}\right]\beta \qquad (4\text{-}53)$$

The factor $\exp(i\omega t)$ vanishes. And, when this expression is divided through by $\exp(i\bar{k}nd)$,

$$-M\omega^2 = \left(e^{i\bar{k}d} + e^{-i\bar{k}d} - 2\right)\beta \qquad (4\text{-}54)$$

This equation will be used to obtain relationships between ω and $\bar{k}$, and, consequently, the ways which these factors affect the traveling waves in the nonhomogeneous string.

The quantity within the parentheses of Equation 4-54 can be simplified in the following way:

$$\left(e^{i\bar{k}d} + e^{-i\bar{k}d} - 2\right) = \left(e^{i\bar{k}d/2} - e^{-i\bar{k}d/2}\right)^2 \qquad (4\text{-}55)$$

Letting $x = \bar{k}d/2$, and using the exponential expression for the sine of x

$$\sin x = \frac{1}{2i}\left(e^{xi} - e^{-xi}\right) \qquad (4\text{-}56)$$

which, when squared becomes

$$\sin^2 x = -\frac{1}{4}\left(e^{xi} - e^{-xi}\right)^2$$

or

$$4\sin^2 x = -\left(e^{i\bar{k}d/2} - e^{-i\bar{k}d/2}\right)^2 = -\left(e^{i\bar{k}d} + e^{-i\bar{k}d} - 2\right) \qquad (4\text{-}57)$$

This is now substituted into Equation 4-54 to obtain

$$M\omega^2 = 4\beta \sin^2 \frac{\bar{k}d}{2}$$

or

$$\omega = \pm\left[\frac{4\beta}{M}\right]^{1/2} \sin\frac{\bar{k}d}{2} \tag{4-58}$$

This is the general expression for the angular frequency as a function of wave vector. Equation 4-58 is now used to obtain maximum values for $\bar{k}$ and ω. Upon differentiation this gives

$$\frac{d\omega}{d\bar{k}} = \left[\frac{4\beta}{M}\right]^{1/2} \frac{d}{2} \cos\frac{\bar{k}d}{2} = 0 \tag{4-59}$$

if a maximum exists. Thus, cos $\bar{k}d/2$ must equal zero, since the other factors do not equal zero. This can only be the case for $\bar{k} = \pi/d$, or

$$\cos\left(\frac{\pi}{d}\cdot\frac{d}{2}\right) = \cos\frac{\pi}{2} = 0$$

So, for this condition

$$\bar{k}_{max} = \pm\frac{\pi}{d} \tag{4-60}$$

This, in turn, can be used to determine ω_{max}, if $\bar{k}_{max}$ is used in Equation 4-58:

$$\omega_{max} = \pm\left[\frac{4\beta}{M}\right]^{1/2} \sin\left(\frac{\pi}{d}\cdot\frac{d}{2}\right) = \pm\left[\frac{4\beta}{M}\right]^{1/2} \tag{4-61a}$$

For ω_{min} Equation 4-58 is equated to zero:

$$\omega_{min} = \pm\left[\frac{4\beta}{M}\right]^{1/2} \sin\left[\frac{\bar{k}d}{2}\right] = 0 \tag{4-61b}$$

Since the coefficient does not equal zero, it follows that

$$\sin\frac{\bar{k}d}{2} = 0 \tag{4-61c}$$

$$\frac{\bar{k}d}{2} = 0, \pi, \text{etc.}$$

$$\bar{k}_{min} = 0, \frac{2\pi}{d}, \text{etc.}$$

Equation 4-58 and the results given by Equations 4-61a and 4-61b permit a description of the behavior of ω as a function of $\bar{k}$ as shown in Figure 4-6. Values of $\bar{k}$ larger in magnitude than $\pm\,\pi/d$ only repeat wave motion already described within that range because of the periodic behavior of Equation 4-58. If a value of $\bar{k} > \pi/d$ is obtained, it is treated by subtracting an integral multiple of $\bar{k} = 2\pi/d$ from it to bring it back within the range $\bar{k}_{max} = \pm\,\pi/d$. Thus, all values of $\bar{k}$ can be represented within this range.

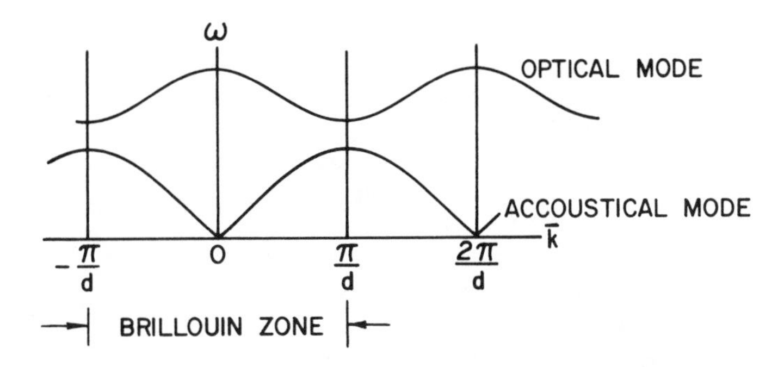

FIGURE 4-6. Variation of angular velocity with wave vector for one dimension.

These waves are waves in which all of the ions are moving in synchronization. This behavior is known as the accoustical mode of vibration and the wave vector range between $\pm\ \pi/d$ is known as the first Brillouin zone. Representations of this kind are commonly used to picture other physical phenomena and are based upon the concept of the reciprocal lattice. The case developed here (shown in Figure 4-6) actually represents a linear reciprocal lattice since the behavior of ω is given in terms of the wave vector which is a function of the reciprocal of the spacing of the ions on the vibrating string.

Another type of vibration may be induced in solids by electromagnetic waves, such as infrared radiation. This is called the optical mode of vibration. Optical modes also can occur when pairs of ions in complex structures move in opposite, rather than the same direction. This type of vibration usually occurs in diatomic lattices. It takes place when the two atomic species vibrate independently of each other. This vibrational mode also lies within the same Brillouin zone as the accoustical mode, but has a different $\omega(\bar{k})$ relationship from that of the accoustical mode (Figure 4-6). While this has interesting physical implications, it will not be discussed here. The accoustical mode is important in many thermal processes in solids.

Consideration will now be given to waves with small values of $\bar{k}$ (those with long wavelengths). For this case, where the sine is small, Equation 4-58 becomes

$$\omega \simeq \pm\left[\frac{4\beta}{M}\right]^{1/2}\frac{\bar{k}d}{2} \tag{4-62}$$

When the factor d/2 is brought under the square root sign

$$\omega \simeq \pm\left[\frac{4}{4}\,\beta d\,\frac{d}{M}\right]^{1/2}\bar{k} \tag{4-63}$$

It will be apparent that Equations 4-47 and 4-44 may now be used here to give

$$\omega \simeq \left[c \cdot \frac{1}{\rho}\right]^{1/2}\bar{k} \tag{4-64}$$

Now, using Equation 4-40, this becomes simply

$$\omega \simeq v\bar{k} \tag{4-65}$$

Thus, where $\bar{k}$ is small, the wavelength is very much greater than the distance between the ions. Here the string of ions can be treated as a homogeneous line. This results from the fact that large numbers of ions are involved in the long wavelengths and such waves are not affected by the discrete structure of the string. It will be shown later that the group and phase velocities are equal in this case.

As the frequency is increased the wavelength shortens and each wave contains fewer numbers of ions. Now the shortest wavelength will be determined from Equation 4-60

$$\bar{k}_{max} = \frac{\pi}{d} = \frac{2\pi}{\lambda_{min}}$$

$$\lambda_{min} = 2d \tag{4-66}$$

The picture of the composition of the short waves under these conditions is quite different from that of the long waves. In both cases the spacing between the oscillating ions remains the same, but here the wavelength is of the same order of magnitude as the ionic spacing. Hence, each high-frequency wave is made up of relatively few ions. In this case the group and phase velocities are no longer equal. Rewriting Equation 4-65 as

$$\omega = v_g\bar{k} \tag{4-67}$$

where v_g is the group velocity. Upon differentiation this becomes

$$\frac{d\omega}{d\bar{k}} = v_g \tag{4-68}$$

Another expression for $d\omega/d\bar{k}$ can be obtained from Equations 4-62, 4-47, and 4-40,

$$\omega = \left[\frac{4\beta}{M}\right]^{1/2} \sin\frac{\bar{k}d}{2} = \left[\frac{4c/d}{\rho d}\right]^{1/2} \sin\frac{\bar{k}d}{2} = \frac{2}{d}\, v \sin\frac{\bar{k}d}{2} \tag{4-69}$$

which upon differentiation gives

$$\frac{d\omega}{d\bar{k}} = \frac{2}{d}\cdot\frac{d}{2} v \cos\frac{\bar{k}d}{2} \tag{4-70}$$

Then equating Equations 4-70 and 4-68, and obtaining

$$v_g = v \cos\frac{\bar{k}d}{2} \tag{4-71}$$

If this relationship is correct, the phase and group velocities should be equal at long wavelengths (small $\bar{k}$). Under these conditions Equation 4-70 becomes

$$v_g = v \cos\frac{\bar{k}d}{2} \simeq v \cos 0 = v \tag{4-72}$$

So Equation 4-72 verifies the statement made earlier: that the long waves composed of many ions have identical phase and group velocities. This permits their treatment as quantized wave packets.

For the shortest possible wavelength, $\bar{k} = \pi/d$ (Equation 4-60). In this limiting case Equation 4-71 becomes

$$v_g = v \cos \frac{\pi}{d} \cdot \frac{d}{2} = 0$$

Here, the group velocity does not exist so "wave packets" are not present.

If v_s is the velocity of sound along the string of ions, then, by Equations 4-65 and 4-69, for $\bar{k}_{max}$

$$v = \frac{\omega}{\bar{k}} = \frac{1}{\bar{k}} \cdot \frac{2}{d} v_s \sin \frac{\bar{k}d}{2} = \frac{d}{\pi} \cdot \frac{2}{d} v_s \sin \frac{\pi}{d} \cdot \frac{d}{2}$$

$$v = \frac{2v_s}{\pi} \sin \frac{\pi}{2} = \frac{2v_s}{\pi}$$

Reintroducing this into Equation 4-65

$$\omega_{max} = v\bar{k}_{max} = \frac{2v_s}{\pi} \cdot \frac{\pi}{d} = \frac{2v_s}{d} \tag{4-73}$$

gives ω_{max} in terms of the velocity of sound along the string of ions. This equation represents the upper limit of Equations 4-28 and 4-79 to be derived in the next section.

4.1.3.4. Application to the Born-von Kármán Model

The behavior of the strings of ions previously derived may now be applied to the Born-von Kármán model of a solid. Specifically, it is necessary to obtain an expression for $N(\omega)d\omega$ of Equation 4-28, the density of vibrational states, or the number of vibrational modes in the range between ω and $\omega + d\omega$. Since ω is a function of $\bar{k}$, it becomes necessary to determine this in terms of $\bar{k}$, the wave vector. This can be obtained starting with

$$N(\omega)d\omega = \frac{dn}{d\bar{k}} \cdot \frac{d\bar{k}}{d\omega} \cdot d\omega \tag{4-74}$$

This is the product of the number of modes per range of $\bar{k}$, the variation of $\bar{k}$ with ω, and the range of ω.

The first of these factors may be obtained with the help of Figure 4-7.

The relationship between the wavelength and the fixed length, L, of the string, in which n is the number of half wavelengths, is

$$\lambda = \frac{2L}{n} \tag{4-75a}$$

Equation 4-75a may be equated to the expression for wavelength as a function of wave vector to get

$$\frac{2L}{n} = \frac{2\pi}{\bar{k}}$$

or

$$\bar{k} = \frac{n\pi}{L} \tag{4-75}$$

Then, differentiating,

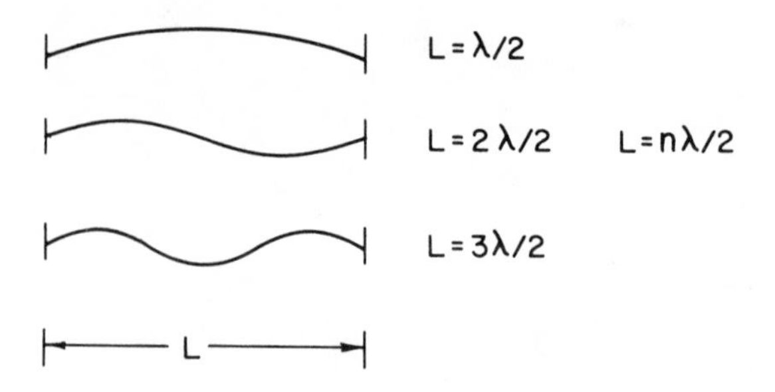

FIGURE 4-7. Waves in a string of fixed length L.

$$\frac{d\bar{k}}{dn} = \frac{\pi}{L} \text{ or } \frac{dn}{d\bar{k}} = \frac{L}{\pi} \tag{4-76}$$

which gives the first factor in Equation 4-74.

The second factor is obtained by starting with Equation 4-58 and substituting 4-61a

$$\omega = \left[\frac{4\beta}{M}\right]^{1/2} \sin\frac{\bar{k}d}{2} = \omega_{max} \sin\frac{\bar{k}d}{2}$$

or

$$\frac{\omega}{\omega_{max}} = \sin\frac{\bar{k}d}{2}$$

$$\frac{\bar{k}d}{2} = \sin^{-1}\frac{\omega}{\omega_{max}}$$

$$\bar{k} = \frac{2}{d}\sin^{-1}\frac{\omega}{\omega_{max}}$$

Then, upon differentiation

$$\frac{d\bar{k}}{d\omega} = \frac{2}{d[\omega_{max}^2 - \omega^2]^{1/2}} \tag{4-77}$$

Equation 4-74 may now be rewritten by substituting Equations 4-76 and 4-77 into it to give the density of vibrational states as

$$N(\omega)d\omega = \frac{L}{\pi} \cdot \frac{2}{d[\omega_{max}^2 - \omega^2]^{1/2}} d\omega \tag{4-78}$$

This may now be used in Equation 4-28, to obtain the Born-von Kármán expression for the internal energy of a solid as

$$U = \frac{2L\hbar}{\pi d}\int_0^{\omega_{max}} \frac{\omega d\omega}{[\exp(\hbar\omega/k_BT) - 1]\,[\omega_{max}^2 - \omega^2]^{1/2}} \tag{4-79}$$

where the upper limit is given by Equation 4-73. The internal energy of a solid obtained in this way is the sum of the energies of parallel linear arrays of coupled ions acting as standing waves. The differentiation of this expression with respect to temperature will give the heat capacity of a solid so constituted.

It is very laborious to determine the heat capacity of real solids using Equation 4-79, because of the difficulties involved with the function for ω. It will be recalled (Equation 4-61a) that the force constants (β_i) for the various directions in solids must be known to determine ω_{max}. This information is rarely available.

A less intricate expression for approximating the internal energy of a solid than that given by Equation 4-79 can be obtained by means of a less sophisticated approximation for $N(\omega)$. This can be treated more readily than that given by Equation 4-78. Starting with Equation 4-75a, for one set of normal modes, and recalling that $\lambda\nu = v_s$,

$$\lambda = \frac{2L}{n} = \frac{v_s}{\nu}$$

Treating v_s as a constant, this is rearranged to give

$$n = \frac{2L}{v_s}\nu \text{ and } dn = \frac{2L}{v_s}d\nu$$

The density of states is obtained as a function of ω by the use of $\nu = \omega/2\pi$. This results in

$$dN(\omega) = \frac{2L}{v_s}\cdot\frac{d\omega}{2\pi} = \frac{L}{\pi v_s}d\omega \tag{4-78a}$$

The substitution of this function into Equation 4-28 gives, for one set of normal modes,

$$U_1 = \frac{L\hbar}{\pi v_s}\int_0^{\omega'_{max}} \frac{\omega d\omega}{\exp(\hbar\omega/k_BT) - 1}$$

When all of the normal modes are taken into account, the internal energy of a solid based upon the Born-von Kârmán model may be approximated by

$$U = \frac{3L\hbar}{\pi v_s}\int_0^{\omega'_{max}} \frac{\omega d\omega}{\exp(\hbar\omega/k_BT) - 1} \tag{4-79a}$$

The upper limit of the integral is obtained by integrating Equation 4-78a and rearranging it to obtain

$$\omega'_{max} = \frac{N\pi v_s}{L}$$

It will be noted that this upper limit is different from that given by Equation 4-73 which is the comparable upper limit for Equation 4-79.

Equation 4-79a is readily treated for high temperatures. Here the first two terms of the series are used for the exponential term in the denominator. The simplified expression is integrated and gives $U \cong 3Nk_BT$; this gives $C_v \cong 3R$, in agreement with the findings of Dulong and Petit.

Equation 4-79a is also used for the low temperature approximation of internal energy. It may be expressed as

$$U = \frac{3L}{\pi v_s}\int_0^{\omega'_{max}} \frac{\hbar\omega d\omega}{\exp(\hbar\omega/k_BT) - 1}$$

$$= \frac{3Lk_BT}{\pi v_s}\int_0^{\omega'_{max}} \frac{\frac{\hbar\omega}{k_BT}d\omega}{\exp(\hbar\omega/k_BT) - 1} \quad (4\text{-}79b)$$

This integral is found, from the tables, to express the internal energy as

$$U = \frac{3Lk_BT}{\pi v_s} \ln \frac{\exp(\hbar\omega/k_BT)}{\exp(\hbar\omega/k_BT) - 1}\Bigg|_0^{\omega'_{max}}$$

The limits are applied after series approximations again are made for the exponential terms. This provides a function which can be manipulated much more readily. The derivative of this function with respect to temperature gives the heat capacity.

The expression for C_v obtained in this way includes a logarithmic term containing k_BT. For temperatures less than about 50 K this term can be approximated as being negligible; this has the effect of eliminating the log term and simplifying the equation. The other term in this equation for C_v contains $(k_BT + 1)$ in its denominator. The same approximation for k_BT results in C_v being a linear function of temperature; C_v approaches zero as T approaches zero. The curve for C_v derived from Equation 4-79 shows the same behavior, but has a slightly smaller slope than that from Equation 4-79b.

Thus, the heat capacity obtained from Equations 4-79a and 4-79b approaches the proper limits for both 0 K and relatively high temperatures. It fails in that it cannot demonstrate that C_v varies as T^3 in the very low range of temperatures. The same comments also apply to C_v derived from Equation 4-79, the Born-von Kármán equation. These failures result from the fact that the interactions of the vibrating ions comprising the solid are much more complex than those capable of description by the oscillations of strings of ions.

4.1.4. The Debye Model

The Debye model (1912) overcomes most of the difficulties inherent in the attempts to describe the heat capacity of solids. Here the solid is treated as an isotropic, continuous, homogeneous medium instead of merely as an assembly of oscillating ions or "strings" of ions. Again, the electron energy is neglected. This solid thus is regarded as being a homogeneous, monatomic substance. Discrete energy states, which are the equivalent of quantized standing waves, can be propagated within the solid. This model treats the solid as though it was pulsating with the standing waves. In this respect such an approach gives a simpler, more realistic representation than the Born-von Kármán model. Here, however, an integration over all of the vibrating ions of the solid as normal modes, rather than a summation over "strings" of ions, is employed to obtain the internal energy of the solid. The limitation of 3N normal modes takes the discrete ionic structure into account.

It will be recalled from prior discussions that it is possible to have any number of normal modes of vibration, depending upon the number of vibrating ions. When complex ionic vibrations exist, they can be represented by the addition of normal modes. In addition, when the phase and group velocities are equal, Equation 4-72, the waves may be quantized and treated as discrete energy states.

The internal energy, U, of the solid is dependent upon the average energy of each quantized standing wave (Equation 4-19) and the total number of such waves between ν and $\nu + d\nu$, known as the "density of states", or $N(\nu)d\nu$. Thus, the total internal energy of a solid, based upon this model is

$$U = \int_{0}^{\nu_{max}} \frac{h\nu}{\exp(h\nu/k_BT) - 1} N(\nu)\,d\nu \qquad (4\text{-}80)$$

Here the upper limit is taken as ν_{max}, but the redundancy which occurs beyond 3N normal vibrational modes must be taken into account. The lower limit is taken as being zero. In actuality it is greater than zero, but for low energies ν is small and only negligible errors are introduced. This will be discussed later.

The density of states must be derived. The total number of standing waves must be equal to 3N. So,

$$\int_{0}^{\nu_{max}} N(\nu)\,d\nu = 3N \qquad (4\text{-}81)$$

It is necessary to count the number of waves in the solid to determine $N(\nu)d\nu$. Consider standing waves in a cube of the solid, the length of whose side is a, as in Figure 4-8. From this it is seen that

$$\lambda/_2 = x \cos \theta_x$$

Since all three dimensions of the solid must be considered, similar expressions must be obtained for the Y and Z directions. In all of these expressions $\cos \theta$ is the direction cosine. Since standing waves are being considered, all of the waves must be such that they consist of an integral number of half wavelengths regardless of the direction of their propagation within the solid. If, for convenience, ℓ_i designates the direction cosines, then, summarized in tabular form:

$$a = \frac{\lambda}{2\ell_x} n_x \qquad \ell_x = \frac{n_x \lambda}{2a} \qquad \ell_x^2 = \frac{n_x^2 \lambda^2}{4a^2}$$

$$a = \frac{\lambda}{2\ell_y} n_y \qquad \ell_y = \frac{n_y \ell}{2a} \qquad \ell_y^2 = \frac{n_y^2 \lambda^2}{4a^2}$$

$$a = \frac{\lambda}{2\ell_z} n_z \qquad \ell_z = \frac{n_z \lambda}{2a} \qquad \ell_z^2 = \frac{n_z^2 \lambda^2}{4a^2}$$

Recalling that the ℓ_i are direction cosines, and that the sum of their squares equals unity, the addition of the squared terms gives

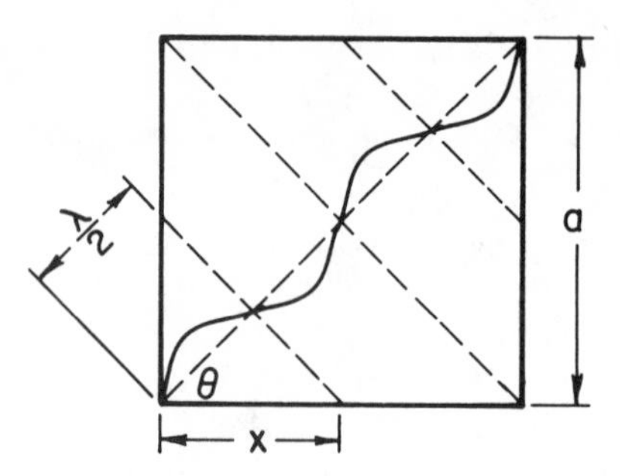

FIGURE 4-8. Schematic diagram of a wave in a solid.

$$\frac{\lambda^2}{4a^2}(n_x^2 + n_y^2 + n_z^2) = 1$$

or

$$\frac{4a^2}{\lambda^2} = n_x^2 + n_y^2 + n_z^2 \quad (4\text{-}82)$$

Thus, the volume of the solid and the waves has been quantized. Each of the vibrational quanta are known as "phonons".

The distance from the origin of any point in the volume of the solid is given by

$$r = [n_x^2 + n_y^2 + n_z^2]^{1/2} = \frac{2a}{\lambda} \quad (4\text{-}83)$$

All points corresponding to the nodes of the standing waves in the solid must be on the surfaces of spheres whose radii are given by Equation 4-83. The velocity of sound is given by

$$c = \lambda\nu \quad (4\text{-}84)$$

When this is substituted into Equations 4-83 for λ

$$r = \frac{2a\nu}{c} \quad (4\text{-}85)$$

Now consider all of the vibrational modes with frequencies between ν and $\nu + d\nu$; $N(\nu)d\nu$ of such modes must be within this range of frequencies. These must be contained within the volume formed between the surfaces of two spheres whose radii are $2a\nu/c$ and $2a(\nu + d\nu)/c$, respectively. Thus, in the first octant (the other octants represent redundant modes),

$$dV = \frac{4\pi r^2 dr}{8} \quad (4\text{-}86)$$

Equation 4-85 and its derivative with respect to frequency are substituted into Equation 4-86 to obtain

$$dV = \frac{\pi}{2} r^2 dr = \frac{\pi}{2}\left[\frac{2a\nu}{c}\right]^2 \cdot \left[\frac{2a}{c} d\nu\right] \quad (4\text{-}87)$$

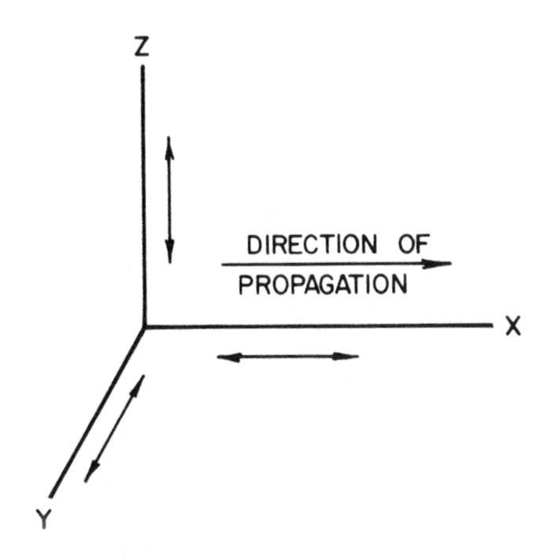

FIGURE 4-9. Possible oscillations of particles within a solid.

The number of modes in dV is just N(ν)dν, so Equation 4-87 becomes

$$N(\nu)d\nu = \frac{4\pi a^3}{c^3}\nu^2 d\nu \tag{4-88}$$

or

$$N(\nu)d\nu = \frac{4\pi V}{c^3}\nu^2 d\nu \tag{4-88a}$$

since $V = a^3$. Equation 4-88 is the expression for the density of states, or the number of quantized waves per unit infinitesimal volume, for the Debye model.

Now consider the types of waves which are present within the solid, with the aid of Figure 4-9. The oscillation directions of the ions are given by the double-headed arrows. The oscillations parallel to both the direction of propagation and to the X axis constitute longitudinal waves. The oscillations of ions parallel to the Y and Z axes are perpendicular to the direction of propagation and constitute transverse waves. Thus, for the longitudinal waves, using Equation 4-88a,

$$N(\nu)_{L_X} = \frac{4\pi V\nu^2}{c^3_{L_X}} \tag{4-88c}$$

and for the transverse waves

$$N(\nu)_{T_Y} = \frac{4\pi V\nu^2}{c^3_{T_Y}} \tag{4-88d}$$

and

$$N(\nu)_{T_Z} = \frac{4\pi V\nu^2}{c^3_{T_Z}} \tag{4-88e}$$

in which c_{L_X}, c_{T_Y} and c_{T_Z} are the velocities of sound in the longitudinal and two transverse directions, respectively.

Since the solid originally was assumed to be a continuous, homogeneous, isotropic medium, it may also be assumed that

$$c_{T_Y} = c_{T_Z} = c_T \tag{4-89}$$

In other words, the velocities of the transverse waves are assumed to be independent of their direction within the lattice. This gives

$$N(\nu) = 4\pi V\nu^2 \left[\frac{1}{c_L^3} + \frac{2}{c_T^3}\right] \tag{4-90}$$

Now, if the further approximation is made, on the same basis as Equation 4-89, that the longitudinal and transverse waves have the same velocities, then

$$c_L = c_T = c \tag{4-91}$$

This approximation simplifies Equation 4-90:

$$N(\nu) = 4\pi V\nu^2 \left[\frac{3}{c^3}\right] \tag{4-92}$$

It will be shown later that the first of these assumptions is reasonable, (Equation 4-89) but the second (Equation 4-91) is poor and limits the accuracy of the model. However, when Equation 4-92 is used, one obtains

$$\int_0^{\nu_{max}} N(\nu)d\nu = \frac{12\pi V}{c^3}\int_0^{\nu_{max}} \nu^2 d\nu = \frac{4\pi V\nu_{max}^3}{c^3} \tag{4-93}$$

Equation 4-93 could present difficulties. In a continuum, ν could vary between 0 and infinity. However, the lower limit of λ (ν_{max}) cannot be less than twice the interionic distance (Equation 4-66) which determines the maximum value for ν. But, the upper limit can only reach some cut-off frequency because no more than 3N normal modes of vibration are possible. Thus, the upper limit must correspond to the 3N modes and no more. Hence, $\nu_{max} = \nu_c$, some cut-off frequency. Therefore,

$$\int_0^{\nu_c} N(\nu)d\nu = 3N \tag{4-94}$$

and

$$\frac{12\pi V}{c^3}\int_0^{\nu_c} \nu^2 d\nu = \frac{4\pi V\nu_c^3}{c^3} = 3N \tag{4-95}$$

From this the cut-off frequency is found to be

$$\nu_c = \left[\frac{3N}{4\pi V}\right]^{1/3} c \tag{4-96}$$

and the minimum wavelength

$$\lambda_{min} = \frac{c}{\nu_c} = \left[\frac{4\pi V}{3N}\right]^{1/3} \qquad (4\text{-}97)$$

Since the fraction V/N in Equation 4-97 is the volume occupied by one of the oscillating ions, λ_{min} cannot equal zero, but must be at least two ionic diameters (Equation 4-66).

The velocity of sound and volume can be eliminated from these expressions. Starting with Equation 4-97, after rearranging, cubing and inverting, it becomes

$$\frac{1}{c^3} = \frac{1}{\nu_c^3} \cdot \frac{3N}{4\pi V} \qquad (4\text{-}98)$$

This may be substituted into Equation 4-92 to get

$$N(\nu) = 12\pi V\nu^2 \cdot \frac{1}{\nu_c^3} \cdot \frac{3N}{4\pi V}$$

or

$$N(\nu) = \frac{9N\nu^2}{\nu_c^3} \qquad (4\text{-}99)$$

This is an approximation of the Debye function for the density of states. When this is applied to Equation 4-80, the Debye expression for the internal energy of a solid is obtained:

$$U = \int_0^{\nu_c} \frac{h\nu}{\exp(h\nu/k_BT) - 1} N(\nu)\,d\nu \qquad (4\text{-}80)$$

which becomes

$$U = \frac{9N}{\nu_c^3} \int_0^{\nu_c} \frac{h\nu^3\,d\nu}{\exp(h\nu/k_BT) - 1} \qquad (4\text{-}100)$$

Before using this equation to determine the heat capacity, Equation 4-100 will be changed to a more tractable form. A temperature, the Debye temperature, is defined as

$$\theta_D = \frac{h\nu_c}{k_B} \qquad (4\text{-}101)$$

where k_B is Boltzmann's constant. The following relationship, derived from Equation 4-101, also will be used:

$$\frac{1}{\theta_D} = \frac{k_B}{h\nu_c}$$

$$\frac{1}{\theta_D^3} = \frac{k_B^3}{h^3 \nu_c^3}$$

and

$$\frac{1}{\nu_c^3} = \frac{h^3}{\theta_D^3 k_B^3}$$

Now let the exponent in Equation 4-100 be

$$x = \frac{h\nu}{k_B T} \tag{4-102}$$

then

$$\nu = \frac{x k_B T}{h}$$

$$\nu^3 = \frac{x^3 k_B^3 T^3}{h^3}$$

Differentiating Equation 4-102 gives

$$dx = \frac{h}{k_B T} d\nu \text{ or } d\nu = \frac{k_B T}{h} dx \tag{4-103}$$

By means of these relations Equation 4-100 is transformed to

$$U = 9N \left[\frac{h^3}{\theta_D^3 k_B^3}\right] \int \frac{h \left[\frac{x^3 k_B^3 T^3}{h^3}\right] \frac{k_B T}{h}}{e^x - 1} dx \tag{4-104}$$

or

$$U = 9N k_B \frac{T^4}{\theta_D^3} \int \frac{x^3 \, dx}{e^x - 1} \tag{4-105}$$

The limits of this integral must be determined. Using Equations 4-102 and 4-101

$$x = \frac{h\nu}{k_B T}$$

$$\theta_D = \frac{h\nu_c}{k_B}$$

$$x_c = \frac{\theta_D}{T} \tag{4-106}$$

which gives the upper limit. As noted previously, the lower limit is approximated by setting it equal to zero, but remembering that it cannot be less than that frequency corresponding to a wavelength equal to the maximum wavelength. Thus, Equation 4-105 becomes

$$U = 9R\frac{T^4}{\theta_D^3}\int_o^{x_c = \theta_D/T} \frac{x^3}{e^x - 1}\,dx \qquad (4\text{-}107)$$

recalling that $Nk_B = R$. The heat capacity of the solid is obtained from the differentiation of Equation 4-107,

$$C_V = \frac{\partial U}{\partial T} = 9R\left\{\frac{4T^3}{\theta_D^3}\int_o^{\theta_D/T} \frac{x^3dx}{e^x - 1} + \frac{T^4}{\theta_D^3}\frac{d}{dT}\left[\int_o^{\theta_D/T} \frac{x^3dx}{e^x - 1}\right]\right\} \qquad (4\text{-}108)$$

Equation 4-108 is used, and the indicated differentiation is performed.

$$\frac{T^4}{\theta_D^3}\frac{d}{dT}\left[\int_o^{\theta_D/T} \frac{x^3dx}{e^x - 1}\right] = \frac{T^4}{\theta_D^3}\frac{d}{dT}\left[\int_o^{\theta_D/T} \frac{\left(\frac{\theta_D}{T}\right)^3\cdot\left(-\frac{\theta_D}{T^2}\right)dT}{e^{\theta_D/T} - 1}\right]$$

$$= \frac{T^4}{\theta_D^3}\cdot\frac{\frac{\theta_D^4}{T^5}(-1)}{e^{\theta_D/T} - 1} = \frac{\theta_D}{T}\cdot\frac{-1}{e^{\theta_D/T} - 1} \qquad (4\text{-}109)$$

Thus,

$$C_V = 9R\left[\frac{4T^3}{\theta_D^3}\int_o^{\theta_D/T} \frac{x^3dx}{e^x - 1} - \frac{\theta_D}{T}\cdot\frac{1}{e^{\theta_D/T} - 1}\right] \qquad (4\text{-}110)$$

which is an approximate, simplified general expression for the heat capacity of a solid given by the Debye model.

If this expression is correct, it should give the classical value at high temperatures. So, for large T the fraction under the integral sign becomes, when a series approximation is used for e^x,

$$\frac{x^3}{e^x - 1} \simeq \frac{x^3}{1 + x - 1} = x^2$$

And, the second term within the brackets becomes

$$\frac{\theta_D}{T}\cdot\frac{1}{e^{\theta_D/T} - 1} \simeq \frac{\theta_D}{T}\cdot\frac{1}{1 + \theta_D/T - 1} = \frac{\theta_D}{T}\cdot\frac{T}{\theta_D} = 1$$

The approximate expression for high temperatures becomes

$$C_V \simeq 9R\left[\frac{4T^3}{\theta_D^3}\int_0^x x^2\,dx - 1\right]$$

When integrated, and θ_D/T is substituted for x,

$$C_V \simeq 9R\left[\frac{4T^3}{\theta_D^3}\cdot\frac{x^3}{3} - 1\right] = 9R\left[\frac{4T^3}{\theta_D^3}\cdot\frac{\theta_D^3}{3T^3} - 1\right]$$

$$C_V \simeq 9R\left[\frac{4}{3} - 1\right] = 3R$$

Thus, the Debye model gives correct results for high temperatures.

Equation 4-110 must be evaluated to determine its behavior at low temperatures. Consider the second term in the brackets of Equation 4-110 as T approaches zero:

$$\lim_{T \to 0}\left[\frac{\theta_D}{T}\cdot\frac{1}{e^{\theta_D/T} - 1}\right] = 0$$

because $\theta_D/T \to \infty$ as $T \to 0$, but $\exp(-\theta_D/T) \to 0$ as $T \to 0$ so the fraction $\to 0$ as $T \to 0$ and their product $\to 0$. The value of the integral in Equation 4-110 is $\pi^4/15$. So,

$$C_V = 9R\left[\frac{4T^3}{\theta_D^3}\,\frac{\pi^4}{15}\right] = \frac{12}{5}\pi^4\left[\frac{T}{\theta_D}\right]^3 R \tag{4-111}$$

This is known as the T^3 law and predicts that the heat capacity of a solid should approach zero as T^3 approaches zero. This agrees with the observed experimental behavior in the temperature range up to about $\theta_D/10$ for many materials. It also removes the difficulty experienced by the Einstein model, where C_V approached zero too quickly. This treatment is applicable only to nonconducting materials since electron effects have been excluded.

The Debye temperature is an important parameter because it is that temperature above which the vibrational energy is large enough to give average ionic displacements ($\bar{x}$) such that

$$\frac{1}{2}K\bar{x}^2 = \text{mean potential energy} = \frac{1}{2}k_BT \tag{4-112}$$

so that classical mechanics can be employed. Below this temperature quantum mechanics must be used. This explains the partial success of the classical approach, Equation 4-10. Further insight into θ_D may be obtained from Equation 4-101. Noting that $\nu_c = v_s/\lambda_c$, and substituting this into the expression for θ_D gives

$$\theta_D = \frac{h\,v_s}{k_B\lambda_c} \tag{4-113}$$

It is apparent that θ_D depends upon the velocity of sound in the solid. This is not a

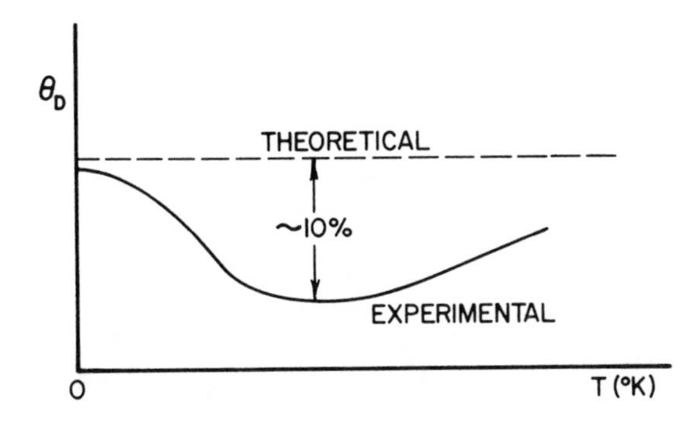

FIGURE 4-10. Variation of the Debye temperature of a metal as a function of temperature; $T \ll \theta_D$.

constant, but is dependent upon such factors as the temperature, the lattice spacing, or its density, and the way in which the phonons, or quanta of lattice vibrational energy, are transmitted and/or scattered by the lattice. In addition, the simplifying assumptions and approximations must also be considered. Thus, the Debye temperature is not exactly a constant; it also varies slightly depending upon the approach used in its experimental determination as well as the range of temperatures in which it is determined. Equation 4-111 provides one way to determine θ_D; it also may be calculated from measurements of electrical resistivity. Typical variations in the Debye temperature are shown in Figure 4-10. The treatment given here is applicable only to nonconductors and to the ion cores of a metallic lattice, since the electron effects are not included. These are discussed in Chapter 5. θ_D actually represents C_V at 96% of its limit when all factors are considered.

The theory treats θ_D as being constant with temperature, as shown in the figure as a broken line. The experimentally determined values for this parameter for copper show a variation of about 10%. Other substances show different ranges of variations, but most appear to have the same general type of variations of θ_D with T as shown in the figure.

The magnitude of θ_D appears to influence the way in which some metals behave mechanically. Both gold and lead have low values for θ_D and have low recrystallization temperatures. Both of these are extremely malleable and recrystallize during working at room temperature. This behavior might be explained by Equation 4-112 on the basis that their ionic displacements in the neighborhood of room temperature are comparable to those of other metals at elevated temperatures, hence their relative ease of working and low recrystallization temperatures.

Einstein (1911) noted a relationship between the characteristic temperature and mechanical behavior, in this case compressibility. This relationship is

$$\theta_E = \frac{13.25 \times 10^{-4}}{A^{1/3} \rho^{1/6} \chi^{1/2}}$$

in which A is the atomic weight, ϱ is the density and χ is the compressibility. An empirical formula by Lindemann also may be used to approximate the Debye temperature:

$$\theta_D = C \left[\frac{T_M}{A\, V^{2/3}} \right]^{1/2}$$

Table 4-1
DEBYE TEMPERATURES AND HEAT CAPACITIES OF SOME ELEMENTS

Element	θ_D (K)[a]	C_p[b]	Element	θ_D (K)[a]	C_p[b]	Element	θ_D (K)[a]	C_p[b]
Li	400	4.95	Mn	400	6.71	Sn (gray)	260	—
Be	1000	5.21	Fe	420	5.30	Sn (white)	170	6.30
B	1250	4.62	Co	385	5.93	Sb	200	6.03
C (diamond)	1860	3.18	Ni	375	6.16	La	132	6.65
Ne	63	4.97	Cu	315	5.86	Pr	74	6.45
Na	150	6.71	Zn	234	6.07	Gd	152	7.03
Mg	318	5.88	Ga	240	6.24	Ta	225	6.43
Al	394	5.82	Ge	360	6.22	W	310	5.97
Si	625	5.91	As	285	5.89	Pt	230	6.19
Ar	85	4.97	Zr	250	6.92	Au	170	6.03
K	100	7.12	Mo	380	5.67	Hg	100	6.50
Ca	230	6.28	Pd	275	6.21	Tl	96	6.29
Ti	400	6.00	Ag	215	5.70	Pb	88	6.12
V	390	5.67	Cd	120	6.19	Bi	120	6.10
Cr	460	5.24	In	129	6.50	Th	100	6.29

[a] Calculated to agree with experimental data in the temperature range at which C_v is approximately half of the value of Dulong and Petit. See Reference 10.
[b] Cal/g-atom/K at 298K. See Reference 11.

where C is a constant (ranging from about 115 to 140), T_M is the melting point in K, and V is the atomic volume.

Some experimental values for θ_D are given in Table 4-1.

4.1.5. Comparison of Models

The same general expression was employed in each theory of heat capacity to obtain a basic relationship for the internal energy of the solid. This was given by

$$U = \int \text{average energy of an oscillator} \times \text{density of states}$$

Each theory employed the same expression (Equation 4-19) for the average energy of an oscillator. Thus, the theories may be compared by examining the difference in the ways the densities of states were counted.

The Einstein model used just one frequency ν_o, for all the independently oscillating ions of the solid. The original Debye approach assumed that the transverse and longitudinal velocities of sound were the same (Equation 4-91). The resultant densities of states of both models are shown graphically in Figure 4-11.

The curve of the density of states for the original Debye model assumes that ν_c, or cut-off frequency, is the same for both longitudinal and transverse waves. The densities of states for the longitudinal and transverse waves, however, are different (Equations 4-88c, d, and e) because $c_T < c_L$. If these differences are taken into account as in Figure 4-12a, then the use of Equation 4-88 can explain the observed behavior somewhat more realistically, but the assumption of a common cut-off frequency for both vibrational modes is incorrect.

The effect of the Born modification may be shown by general form of Equation 4-88:

$$N(\nu) = \text{const}\ \frac{\nu^2}{c^3} \tag{4-114}$$

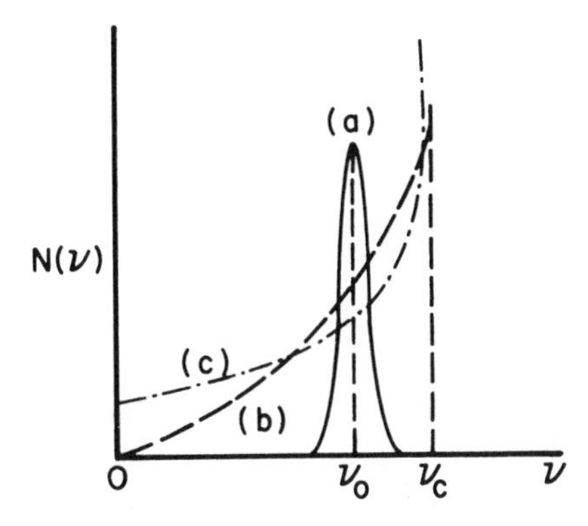

FIGURE 4-11. Densities of states for (a) the Einstein Model; (b) the original Debye model; and (c) the Born-von Kármán model. Note its singularity at $\omega = 2\pi\nu_{max}$ (Equation 4-78).

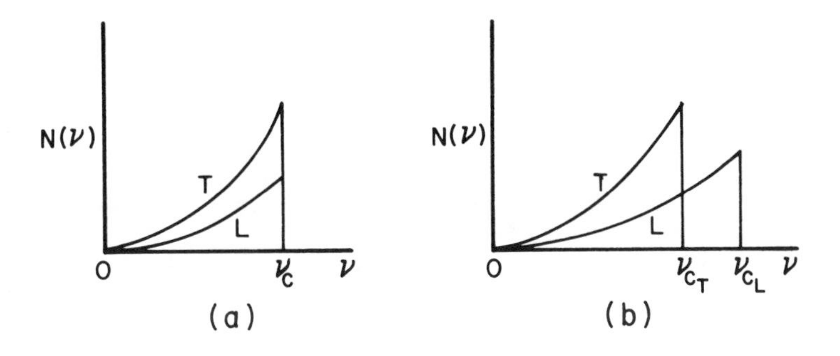

FIGURE 4-12. Modified densities of states. (a) Debye; (b) Born. (From Dekker, A. J., *Solid State Physics,* Prentice-Hall, Englewood Cliffs, N.J., 1957, 42. With permission.)

A minimum wavelength is assumed common to each vibrational mode. This results in a different cut-off frequency for each mode, since $c_T<c_L$, and each gives a different $N(\nu)_T$ and $N(\nu)_L$ (Equation 4-114 and Figure 4-12b). Obviously, the original Debye approximation for $N(\nu)$, based on $c_T = c_L$, is reasonably good only when the theoretical density of states lies between the two distributions shown in the figure, and the cut-off frequency falls between those of each type of wave.

It will be recalled that $N(\nu)$ was determined from the way in which the standing waves fit into the solid cube. In addition, it was previously noted in this section and in Equation 4-66 that λ cannot be less than two interionic spacings. Thus, this minimum wavelength sets the maximum frequency limit. The actual cut-off frequency for each type of vibration is determined by 3N normal modes, respectively.

If the Born modification (Figure 4-12b) is used, where the cut-off frequencies for transverse and longitudinal waves are taken into account, the model, while still oversimplified, provides a closer representation of observed behavior than does the unmodified Debye model.

It must also be recalled that the Debye model assumed an ideal solid. Impurities and imperfections exist in even the best single-crystal specimens. In polycrystalline materials, the deviation from the ideal is compounded by grain boundaries which add further complications. Further, if the solid is a metal, the role of the valence, or conduction, electrons has been neglected. This factor must be considered at low temperatures. Nonetheless, the Debye approximation provides the best model for the behavior of the heat capacity of lattices at low temperatures. It frequently is employed along with the electron contribution, as described in Section 5.6.1, Chapter 5, for the determination of properties of metals at very low temperatures.

4.2. ZERO POINT ENERGY

Classical physicists considered that the ions in a solid were motionless, or ''frozen in'', at 0 K. It is now known that this is not so. Solutions to Schrödinger's equation (Section 3.10, Chapter 3) can be used to demonstrate that the ionic vibrations in a solid at absolute zero are such that appreciable energy is involved.

It was shown that the energy of an oscillator was given by

$$E_n = \left(n + \frac{1}{2}\right) h\nu \tag{3-72}$$

This expression must be used because the classical equation for the energy of an oscillating ion is invalid below θ_D. The ground state, or lowest energy state, of an oscillating ion, with a frequency ν, is

$$E_o = \frac{1}{2} h\nu \tag{4-115}$$

The zero point energy of a solid then is, summing over all of the ions,

$$U_o = \int_o^{\nu_c} \frac{1}{2} h\nu N(\nu)\, d\nu \tag{4-116}$$

Using the original Debye approximation (Equation 4-99),

$$U_o = \int_o^{\nu_c} \frac{1}{2} h\nu \cdot \frac{9N}{\nu_c^3} \nu^2\, d\nu$$

or

$$U_o = \frac{9}{2} \frac{Nh}{\nu_c^3} \int_o^{\nu_c} \nu^3 d\nu = \frac{9}{8} Nh\nu_c \tag{4-117}$$

Since, by Equation 4-101, $h\nu_c = k_B\theta_D$, the least possible internal energy of the solid is approximately

$$U_o = \frac{9}{8} Nk_B\theta_D = \frac{9}{8} R\theta_D \tag{4-118}$$

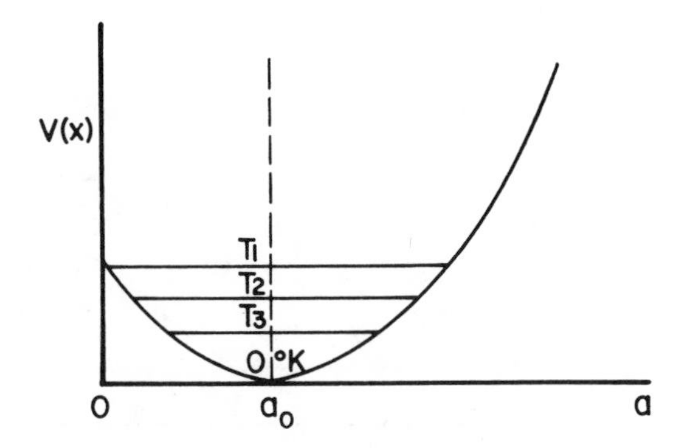

FIGURE 4-13. Average amplitudes of an oscillating particle as a function of temperature ($T_1 > T_2 > T_3 > 0$ K).

This is the same order of magnitude as the energy of an ideal gas (3/2 RT) in the neighborhood of room temperatures, since θ_D for many elements is about 300 K.

On the basis of classical mechanics it would have been expected that the lattice of the solid would be perfect and that all of the ions would have been at rest on their lattice sites. That this is not the case is apparent from Equation 4-118, the zero-point energy. Thus, the internal energy of a solid at 0 K is quite high. There is no state in which the internal energy of a solid is zero.

Since energy is a relative quantity, the lowest energy of a solid is taken to represent that state in which both all of the ions and all of the electrons of the solid attain a unique lowest energy state at 0 K. For certain thermodynamic purposes such a condition may be assigned as being the zero of energy. This is not at variance with the thermodynamics. When the zero-point energy is used as a reference for energy measurements, differences, or changes, in energy are actually being measured and the zero-point energy cancels out. In the same way, the entropy of a substance is arbitrarily designated as being equal to zero for this set of conditions at 0 K, where all the ions and electrons are in the lowest possible entropy state.

4.3. THERMAL EXPANSION

All solids show dimensional changes as the temperature varies. Most solids expand as a function of increasing temperature. This behavior can be explained by considering the vibrations of the ions comprising the solid. As the temperature increases, the amplitudes of the vibrations of the ions become greater and the volume of the solid increases.

If the ions are treated as classical, simple-harmonic oscillators, then the condition sketched in Figure 4-13 should prevail. According to classical ideas, the particles would be at rest at a_o at 0 K. As the temperature increased the particles would oscillate about a_o, their mean, or equilibrium, position; the average amplitude of this oscillation would be the same for all directions. As many ions could be found with positions greater than a_o as those with positions less than a_o. Thus, no thermal expansion can be predicted by the classical approach since the curve is symmetric about a_o. Thus, the curve of Figure 4-13 does not represent the observed conditions.

In order that the observed behavior be accounted for, there must be a net number of ionic oscillations such that the average ion oscillation is greater than a_o at a given temperature. An exaggerated curve deduced from observed behavior is shown in Figure 4-14. This curve must be asymmetric in order to describe the observed expansion. The

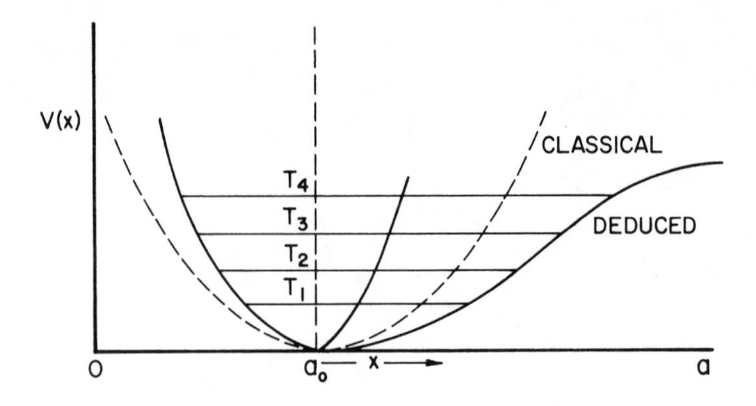

FIGURE 4-14. Average amplitudes of an oscillating particle as a function of temperature ($T_4 > T_3 > T_2 > T_1 > 0$ K).

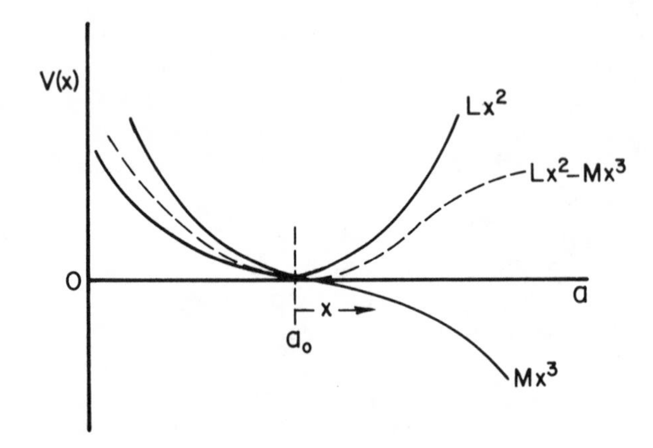

FIGURE 4-15. Schematic diagram for the approximation of the average amplitude of an oscillating particle.

average lattice position of an oscillating ion as a function of temperature is given by the line connecting the midpoints of the ranges of vibration for the four temperatures shown on the deduced curve.

The classical expression, $V(x) = \frac{1}{2}Kx^2$, cannot describe the above behavior, since it is symmetric about a_o. A first approximation of the actual behavior may be obtained by the use of an anharmonic, or repulsive, term to obtain another classical expression which describes the thermal expansion more realistically:

$$V(x) = Lx^2 - Mx^3 \tag{4-119}$$

where x is the displacement from a_o, and the coefficients are related to the bonding energies of the ions. The effect of this anharmonic term is shown schematically as in Figure 4-15.

The average value of the displacement ($\bar{x}$) from a_o is given by

$$\bar{x} = \frac{\int_{-\infty}^{\infty} x \exp\left[-V(x)/k_B T\right] dx}{\int_{-\infty}^{\infty} \exp\left[-V(x)/k_B T\right] dx} \tag{4-120}$$

This will be simplified. First consider the numerator. This is

$$\int_{-\infty}^{\infty} xe^{-V(x)/k_BT}dx = \int_{-\infty}^{\infty} xe^{-(Lx^2 - Mx^3)/k_BT}dx \quad (4\text{-}121)$$

$$= \int_{-\infty}^{\infty} \left[xe^{-Lx^2/k_BT} \cdot e^{Mx^3/k_BT}\right] dx \quad (4\text{-}122)$$

The second exponential factor in Equation 4-122 may be expressed as the first two terms of a series, and

$$= \int_{-\infty}^{\infty} xe^{-Lx^2/k_BT} \left(1 + \frac{Mx^3}{k_BT}\right) dx \quad (4\text{-}123)$$

or, expanding

$$= \int_{-\infty}^{\infty} xe^{-Lx^2/k_BT} dx + \frac{M}{k_BT} \int_{-\infty}^{\infty} x^4e^{-Lx^2/k_BT} dx \quad (4\text{-}124)$$

The first term in Equation 4-124 may be written as

$$\int_{-\infty}^{\infty} xe^{-Lx^2k_BT} dx = 2\int_{o}^{\infty} xe^{-Lx^2/k_BT} dx \quad (4\text{-}125)$$

because it is a symmetrical function. This means that when Equation 4-125 is used in Equation 4-120, the positive and negative portions of the integral will cancel each other. The second integral in Equation 4-124 is given by the general expression

$$\int_{o}^{\infty} x^{2n}e^{-ax^2} dx = \frac{1 \cdot 3 \cdot 5 \cdot\cdot\cdot (2n - 1)}{2^{n+1} a^n} \left[\frac{\pi}{a}\right]^{1/2}$$

So, for $n = 2$ and $a = L/k_BT$,

$$\frac{2M}{k_BT} \int_{-\infty}^{\infty} x^4e^{-Lx^2/k_BT} dx = \frac{M}{k_BT} \left[\frac{3}{4}\left(\frac{k_BT}{L}\right)^{3/2} \pi^{1/2}\right] \quad (4\text{-}126)$$

Consider now the denominator of Equation 4-120. Using only the first term in V(x), it is found, from tables, that

$$\int_{-\infty}^{\infty} e^{-V(x)/k_BT} dx \approx \int_{-\infty}^{\infty} e^{-Lx^2/k_BT} dx = \left[\frac{\pi k_BT}{L}\right]^{1/2} \quad (4\text{-}127)$$

when the anharmonic term is neglected. Equations 4-126 and 4-127 are substituted into Equation 4-120 to give

Table 4-2
SOME SELECTED TEMPERATURE COEFFICIENTS OF THERMAL EXPANSION FOR ELEMENTS NEAR 20°C (CM/CM/°C × 10^6)

Element	α_L	Element	α_L
Aluminum	23.6[a]	Platinum	8.9
Antimony	8.5—10.8[b]	Rhenium	6.7[h]
Beryllium	11.6[c]	Rhodium	8.3
Bismuth	13.3	Ruthenium	9.1
Cadmium	29.8	Selenium	37
Carbon (graphite)	0.6—4.3[a]	Silicon	2.8—7.3
Chromium	6.2	Silver	19.68[i]
Cobalt	13.8	Tantalum	6.5
Copper	16.5	Tellurium	16.75
Gold	14.2	Thorium	12.5[j]
Hafnium	519[d]	Tin	23[i]
Indium	33	Titanium	8.4
Iron	11.76[e]	Tungsten	4.6
Lead	29.3[f]	Uranium	6.8—14.1
Magnesium	27.1[g]	Vanadium	8.3[k]
Molybdenum	4.9[a]	Zinc	39.7[l]
Nickel	13.3[c]	Zirconium (α)	5.88
Palladium	11.76		

Note: Table adapted from Reference 14, Chapter 5.

[a] 20—100°C.
[b] 20—60°C.
[c] 25—100°C.
[d] 20—200°C.
[e] 25°C.
[f] 17—100°C.
[g] in basal plane, 24.3 in c direction.
[h] 20—500°C.
[i] 0—100°C.
[j] 25—1000°C.
[k] 23—100°C.
[l] 20—250°C.

$$\bar{x} \cong \frac{\frac{3}{4} M \frac{(k_B T)^{3/2}}{L^{5/2}} \pi^{1/2}}{\frac{(k_B T)^{1/2}}{L^{1/2}} \pi^{1/2}} = \frac{3}{4} M \frac{k_B T}{L^2} \tag{4-128}$$

Equation 4-128 also may be written as

$$\bar{x} = \frac{3}{4} \frac{M}{L^2} k_B T = \frac{3}{4} \frac{M}{L^2} \bar{U} \tag{4-129}$$

where $\overline{U}$ is the average energy. Using Equation 4-27 this becomes

$$\bar{x} = \frac{3}{4} \frac{M}{L^2} \frac{\hbar\omega}{\exp(\hbar\omega/k_B T) - 1} \tag{4-130}$$

At temperatures above θ_D this can be approximated to be the same as Equation 4-128. Upon differentiation with respect to temperature,

Table 4-3
TYPICAL THERMAL PROPERTIES OF SOME SELECTED COMMERCIAL ALLOYS NEAR ROOM TEMPERATURE

Material	Coefficient of linear expansion (%/°C × 10⁴)	Thermal conductivity (K cal/sec m² °C)
SAE 1020	12.2	0.23
Gray cast iron	12.1	0.20
304 Type stainless steel	17.3	0.08
3003 Aluminum alloy	23.2	0.68
380 Aluminum alloy	20.1	0.42
Copper	16.7	1.70
Yellow brass	18.9	0.52
Aluminum bronze	16.6	0.31
Copper-beryllium	16.7	0.05
Solder (50 Pb-50 Sn)	23.6	0.20
Constantan	14.6	0.10
Titanium (commercial pure)	8.8	0.08

From *Handbook of Chemistry and Physics*, 56th ed., Weast, Robert C., Ed., CRC Press, Cleveland, Ohio, 1975, D171. With permission.

$$\frac{\partial \bar{x}}{\partial T} = \frac{3}{4}\frac{M}{L^2} k_B \tag{4-131}$$

Now, using the original equilibrium interionic distance, a_o, as a reference, and dividing through by it, gives the linear temperature coefficient of expansion as

$$\alpha_L = \frac{\partial \bar{x}}{a_o \partial T} = \frac{3}{4}\frac{M}{L^2}\frac{k_B}{a_o} \tag{4-132}$$

The approximation given by Equation 4-132 does not expressly show α_L to be a function of temperature. However, it will be recalled that the coefficients L and M are related to the bonding energies; these are functions of temperature. The linear coefficient of expansion is, in fact, a small, approximately linear function of temperature above θ_D. This equation gives a reasonable approximation (see Tables 4-2 and 4-3).

For temperatures below θ_D, Equation 4-130 can be approximated by

$$\bar{x} = \frac{3}{4}\frac{M}{L^2} \hbar\omega \exp(-\hbar\omega/k_B T) \tag{4-133}$$

Following the same procedure as before

$$\alpha_L = \frac{\partial \bar{x}}{a_o \partial T} = \frac{3}{4}\frac{Mk_B}{L^2 a_o}\left[\frac{\hbar\omega}{k_B T}\right]^2 \exp(-\hbar\omega/k_B T) \tag{4-134}$$

Thus, α_L will approach zero exponentially as T approaches zero in much the same way as the Einstein expression for C_v (Equation 4-25). This approximation is incorrect for low T, where α_L should vary in the same way as C_v, namely as T^3.

It can be shown, using a thermodynamic approach based upon the Helmoholtz free energy, that a more accurate expression is given by

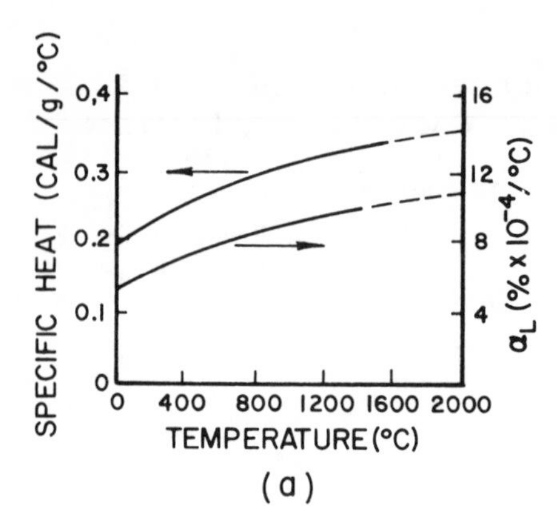

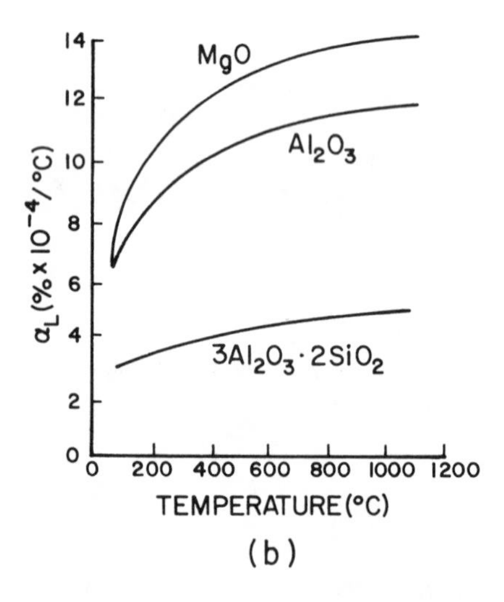

FIGURE 4-16. (a) Parallel behavior of specific heat and temperature coefficient of linear expansion of Al_2O_3; (b) Temperature coefficient of linear expansion of some insulators. (After Kingery, W. D., *Introduction to Ceramics*, John Wiley & Sons, New York, 1960, 470. With permission.)

$$\alpha_L = \frac{\gamma C_V}{3BV} \tag{4-135}$$

where γ is the Grüneisen constant and is defined as

$$\gamma \equiv -\frac{d \ln \theta_D}{d \ln V} = -\frac{V}{\theta_D}\frac{d\theta_D}{dV} \tag{4-136}$$

and B is the bulk modulus (the reciprocal of compressibility) which is given by

$$B = -\frac{VdP}{dV} \tag{4-137}$$

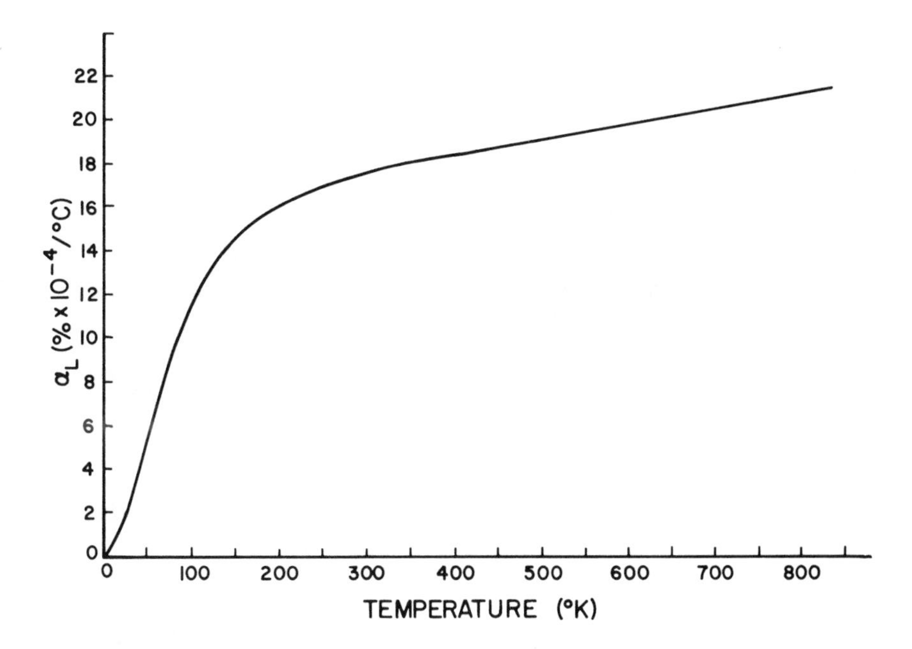

FIGURE 4-17. Temperature coefficient of linear expansion of copper. (After Hahn, T. A., *J. Appl. Phys.*, 41, 5096, 1970. With permission.)

The Grüneisen constant may be closely approximated as being independent of temperature for materials with cubic lattices. Its values vary between 1.3 and 2.5 for metallic elements and halide salts. For purposes of approximation, $\gamma \sim 2$ can be used. The bulk modulus may be calculated from the mechanical properties of materials. This is given by

$$B = \frac{E_y}{3(1 - 2\mu)} \tag{4-138}$$

in which E_y is Young's modulus and μ is Poisson's ratio. Both of these mechanical properties are functions of temperature. Thus, as expected from Equation 4-137 as well, B is a function of temperature.

C_v is the most important factor in Equation 4-135. Thus, the curve of α_L as a function of temperature should be expected to be very similar to that of C_v. This is to be expected since both properties result from the way in which the oscillations of the ions are affected by temperature. Many materials approximate this behavior (Figure 4-16).

Metals show similar behaviors. Since the models used here are based upon isotropic, homogeneous, monatomic solids in which electron effects were omitted, deviations from these ideal conditions would be expected to influence the coefficients of expansion of metals and alloys. Electron effects, including magnetic changes, impurities, imperfections, grain boundaries, internal stresses, ordering, and the presence of other phases all affect α_L just as they do C_v. The behavior of pure copper is shown in Figure 4-17.

In very pure, nearly perfect, single crystals of metals which have cubic lattices, anisotropic variations in α_L are very apparent, the values being identical along the cube edges. Noncubic crystals show pronounced differences in α_L for each principle lattice direction. This results in larger anisotropic effects than usually are shown by cubic lattices. The data in Table 4.2, with one exception, are for polycrystalline elements.

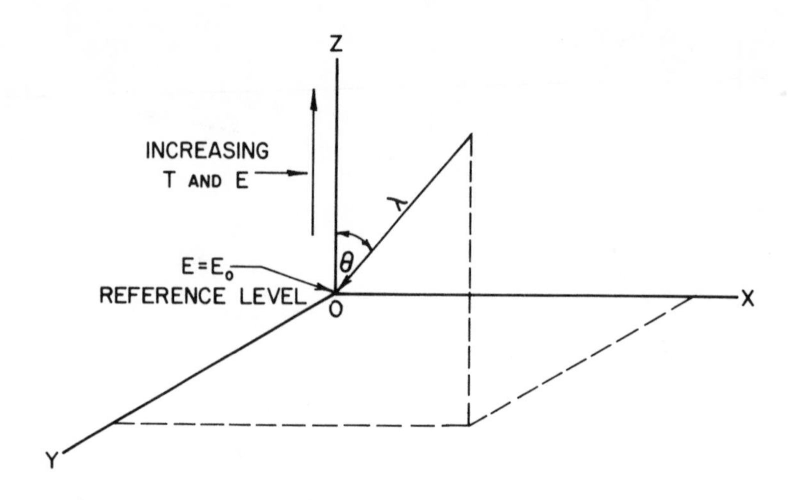

FIGURE 4-18. Schematic diagram of a phonon of wavelength λ interacting with the X-Y plane.

These data represent average values of α_L, in most cases, because of the random orientations of the grains.

4.4. THERMAL CONDUCTIVITY

In this discussion of the thermal conductivity of insulators, the transfer of thermal energy is the result of the ionic vibrations in the solid. The amplitudes of these vibrations increase as a function of temperature. An ion at the surface, vibrating about its equilibrium position at an amplitude corresponding to that of the given temperature, will increase its amplitude when the ambient temperature is increased. This causes periodic forces to act upon adjacent, internal neighbors; their amplitudes increase and the energy is transferred by the phonons. This mechanism continues throughout the entire solid, conducting the heat through it. A continuous flow of heat will occur in this way until thermal equilibrium is reached. When a temperature difference exists across a solid, the vibrational amplitudes of the ions gradually diminish across the solid, from the hotter to the cooler surface; the flow of energy will continue as long as the temperature difference exists.

Consider the transfer of energy by phonons crossing the XY plane (Figure 4-18) which is at energy level E_o. Each phonon arriving at the plane will have an average mean free path λ, make a solid angle θ with the Z axis, and have a mean energy

$$E_0 + \lambda \cos\theta \frac{\partial E}{\partial Z} \tag{4-139}$$

The total energy added by all phonons crossing the plane is

$$\Delta E = \lambda \cos\theta \frac{\partial E}{\partial Z} \times \text{flux} \tag{4-140}$$

The flux, or the number of phonons which cross unit area per unit time, within the solid angle between θ and $\theta + d\theta$, is determined with the help of Figure 4-19. Assume that N phonons each with an average velocity v pass across the surface of the hemisphere in unit time. The area of the hemisphere of unit radius is 2π. The circumference

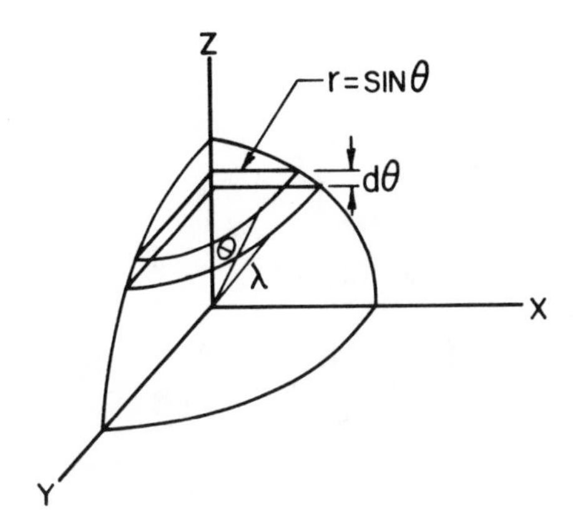

FIGURE 4-19. Sketch for the determination of phonon flux showing only one quarter of the hemisphere of interest.

of the element of area is $2\pi r = 2\pi \sin \theta$. The area of this element is $2\pi \sin \theta d\theta$. The number of phonons which intercept this element of surface is then dN. The ratio of dN/N will equal the ratio of the areas involved. Thus

$$\frac{dN}{N} = \frac{2\pi \sin \theta d\theta}{2\pi}$$

or

$$dN = N \sin \theta d\theta \tag{4-141}$$

The component of the velocity of the phonons parallel to the Z axis is: $v_z = v \cos \theta$. The flux is given by

$$df_Z = dNv_Z \tag{4-142}$$

The flux parallel to the Z axis (perpendicular to the XY plane) is

$$df_Z = dNv \cos \theta \tag{4-143}$$

Or, using Equation 4-141, the expression for dN,

$$df_Z = Nv \sin \theta \cos \theta d\theta \tag{4-144}$$

Now, assuming that as many phonons travel in the positive direction as in the negative, then

$$df_Z = \frac{N}{2} v \sin \theta \cos \theta d\theta \tag{4-145}$$

Referring back to Equation 4-140

$$dE = \lambda \cos \theta \frac{\partial E}{\partial Z} df_Z$$

or, substituting and integrating

$$\Delta E = \frac{1}{2} N v \lambda \frac{\partial E}{\partial Z} \int_o^\pi \cos^2 \theta \sin \theta d\theta \qquad (4\text{-}146)$$

The integral is integrated by parts, as follows:

$$\int_o^\pi \cos^2 \theta \sin \theta d\theta = -\frac{1}{2} \cos^2 \theta - \frac{1}{2} \int_o^\pi \cos^2 \theta \sin \theta d\theta$$

Then transposing

$$\frac{3}{2} \int_o^\pi \cos^2 \theta \sin \theta d\theta = -\frac{1}{2} \left[\cos^3 \theta\right]_o^\pi$$

Clearing and changing the limits

$$2\int_o^{\pi/2} \cos^2 \theta \sin \theta d\theta = -\frac{2}{3} \left[\cos^3 \theta\right]_o^{\pi/2}$$

$$= -\frac{2}{3} [0 - 1] = \frac{2}{3}$$

This result substituted into Equation 4-146 gives the increase in energy at the plane as

$$\Delta E = \frac{1}{3} N v \lambda \frac{\partial E}{\partial Z} \qquad (4\text{-}147)$$

This provides the basis for the determination of the expression for the thermal conductivity of nonconductors.

The thermal energy transferred between two planes perpendicular to the Z axis at temperatures T_2 and T_1, respectively, where $T_2 > T_1$, respectively, is

$$\Delta E = \kappa \frac{T_2 - T_1}{Z_2 - Z_1}$$

where the constant of proportionality, κ, is the thermal conductivity. Or, in different form

$$\Delta E = \kappa \frac{\partial T}{\partial Z} ; \kappa = \Delta E \frac{\partial Z}{\partial T} \qquad (4\text{-}148)$$

Now Equations 4-147 and 4-148 may be equated

$$\Delta E = \frac{1}{3} N v \lambda \frac{\partial E}{\partial Z} = \kappa \frac{\partial T}{\partial Z} \qquad (4\text{-}149)$$

This may be simplified by reexpressing two of the factors in Equation 4-147 as

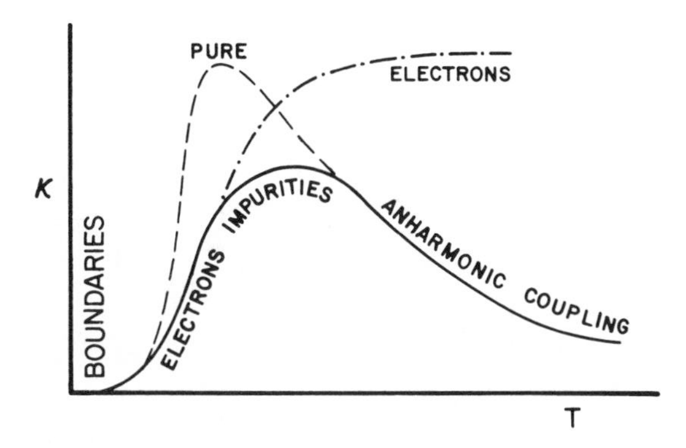

FIGURE 4-20. Theoretical general form of thermal conductivity. The broken curve shows the behavior of a pure crystalline insulator. The dot-dash curve is for that of a metal if only electrons are scattered by phonons. (After Makinson, R. E. B., *Proc. Camb. Philos. Soc.*, 34, 474, 1938. With permission.)

$$N \frac{\partial E}{\partial Z} = N \frac{\partial E}{\partial T} \cdot \frac{\partial T}{\partial Z}$$

and noting that

$$N \frac{\partial E}{\partial T} = C_V$$

Thus, Equation 4-149 becomes

$$\Delta E = \frac{1}{3} C_V \, v \, \lambda \frac{\partial T}{\partial Z} = \kappa \frac{\partial T}{\partial Z} \qquad (4\text{-}150)$$

By inspection of Equation 4-150 it is seen that

$$\kappa = \frac{1}{3} C_V \, v \, \lambda \qquad (4\text{-}151)$$

The thermal conductivity of solids varies widely as a function of temperature. It is apparent from prior discussions that each of the three factors in Equation 4-151 is related to the other. Nevertheless, it appears that each factor assumes a different degree of importance in various ranges of temperature.

Consideration will first be given to the range of temperatures with θ_D as the lower limit. At these temperatures the number of phonons, n, is high. The mean free path of the phonons would be expected to vary inversely as their number; thus, λ would be expected to be small. The larger n becomes, the shorter λ becomes. In addition, as n becomes larger, the greater is the probability of a phonon interacting with other phonons and being scattered. This is known as an "umklapp" (flip-over) process. This is discussed below. Such interactions can result in complex coupling, or anharmonic effects, which effectively further tend to decrease λ. Under such conditions the absorbed thermal energy is transmitted very inefficiently.

In the case of metals, this very low efficiency of energy transmission by the ions of the lattice at high temperatures is the reason for the major role of the electrons in the

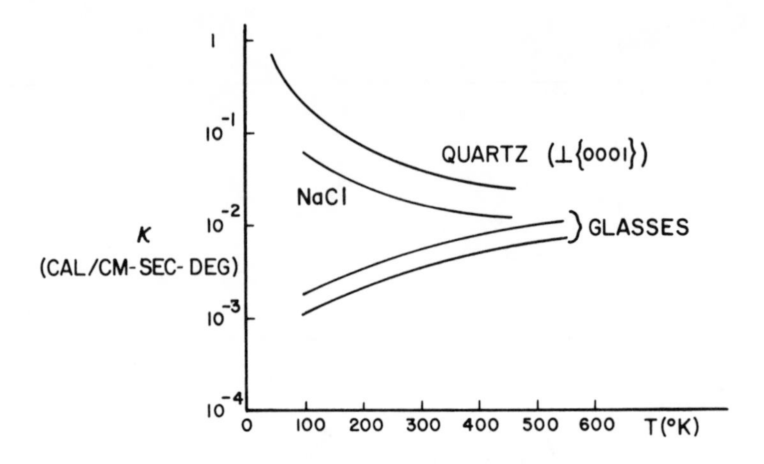

FIGURE 4-21. Thermal conductivities of some nonconductors as functions of temperature. (After Kittel, C., *Introduction to Solid State Physics*, 3rd ed., John Wiley & Sons, New York, 1966, 193. With permission.)

transport of thermal energy. As will be shown subsequently, electrons are not scattered by phonons to the extent that phonons are. Metals are good thermal conductors because of the relative freedom of their valence electrons. In this range the thermal energy transport afforded by the electrons is about fifty times that of the phonons. Therefore, in the range of temperatures above θ_D, the ionic contribution to κ is low because λ is small. This is shown at the high-temperature end of the curve in Figure 4-20. Here the solid line gives the theoretical thermal conductivity of a nonconductor as a function of temperature. The role of electrons in conductors is shown for comparison.

In the low range of temperatures, the factor C_V is small and λ is the predominating factor. In this range λ can become quite large. Here, λ appears to be limited by the dimensions of the solid, and by both internal and surface imperfections. Under these conditions it is possible that a very pure, nearly perfect crystal of an insulator could have a significantly higher thermal conductivity than a polycrystalline metallic material. The effects of imperfections at low temperatures are shown in Figure 4-21. The high contrast between the properties of good crystals and glasses is striking. The glasses are super-cooled liquids. As such, their ''structures'' are highly irregular. Instead of a good crystalline array, as in the quartz or sodium chloride crystals, the glassy state is probably best described by nonuniform, warped, ionic arrays more like those of liquids than of crystals. This absence of uniform ionic structure strongly affects λ, the predominating factor in this range. The more usual lattice imperfections and impurities in crystals affect the wavelength in a similar way, but to a much lesser degree than in glasses. An approximate relationship may be obtained for the behavior of κ at low temperatures, based upon Equation 4-151. According to Equation 4-111, C_V will vary as $(T/\theta_D)^3$, so it will be small. The phonon velocity, v, may be approximated as being constant. The phonon wavelength will be affected by the way in which it interacts with other phonons. This will vary inversely as the number of phonons present. A minimum phonon energy of $k_B\theta_D/2$ is required for such interactions, according to Peierls. The number of such phonons is approximated by Boltzman statistics as exp $[-k_B\theta_D/2k_BT]$ or exp $[-\theta D/2T]$. Since λ is the reciprocal of this, the thermal conductivity should vary as exp $[\theta_D/2T]$. But, κ is expected to decrease exponentially in the low temperature range in agreement with the behavior shown in Figure 4-20.

The thermal conductivity does not become infinite as T approaches zero. In this temperature range the factor $(T/\theta_D)^3$ in the Debye expression (Equation 4-111) ap-

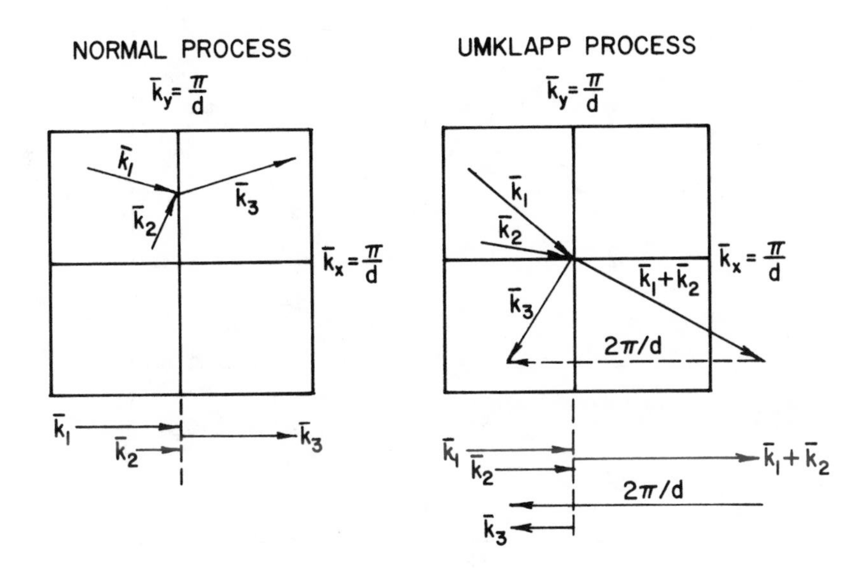

FIGURE 4-22. Brillouin zone treatment of phonon vectors, in two dimensions, with projections of the phonon vectors parallel to $\bar{k}_x$.

proaches zero as T approaches zero. Thus, the thermal conductivity approaches zero even though λ becomes very large.

In the approximate range from $\theta_D/10$ to about $\theta_D/2$, impurities and lattice imperfections play an increasingly important role in thermal conduction. In this temperature range C_V may be approximated as being a linear function of T/θ_D. Here the phonon velocity will vary approximately as $T^{1/2}$. The wavelength may be considered to vary as θ_D/T. Thus, the approximate behavior is

$$\kappa \propto \frac{T}{\theta_D} \cdot T^{1/2} \cdot \frac{\theta_D}{T} = T^{1/2} \tag{4-152}$$

in this temperature range.

At more elevated temperatures, using the neighborhood of θ_D as a lower limit, C_V is essentially constant. The phonon velocity varies approximately as $T^{1/2}$ and λ is roughly proportional to $1/T$. These behaviors give the approximation

$$\kappa \propto 3R \cdot T^{1/2} \cdot \frac{1}{T} = \frac{\text{const}}{T^{1/2}} \tag{4-153}$$

This diminishing value of the thermal conductivity with temperature agrees with the behavior shown in Figure 4-20. It results primarily from phonon-phonon interactions which scatter the phonons. The greater the scattering, the less efficient are the phonons in transmitting the energy in the solid.

Such behavior can be explained in terms of the basis of Figure 4-6. This figure gives a one-dimensional Brillouin zone. It is obvious that three-dimensional zones must be employed for the case of solids. For equally obvious reasons two-dimensional zones will be employed to illustrate this effect in Figure 4-22. Here, normal processes are those in which the resultant of the interaction of phonons lies within the Brillouin zone. Momentum is unchanged when such phonons interact and the energy flux passes through the solid with relative efficiency and with a minimum of decay. Thus, where

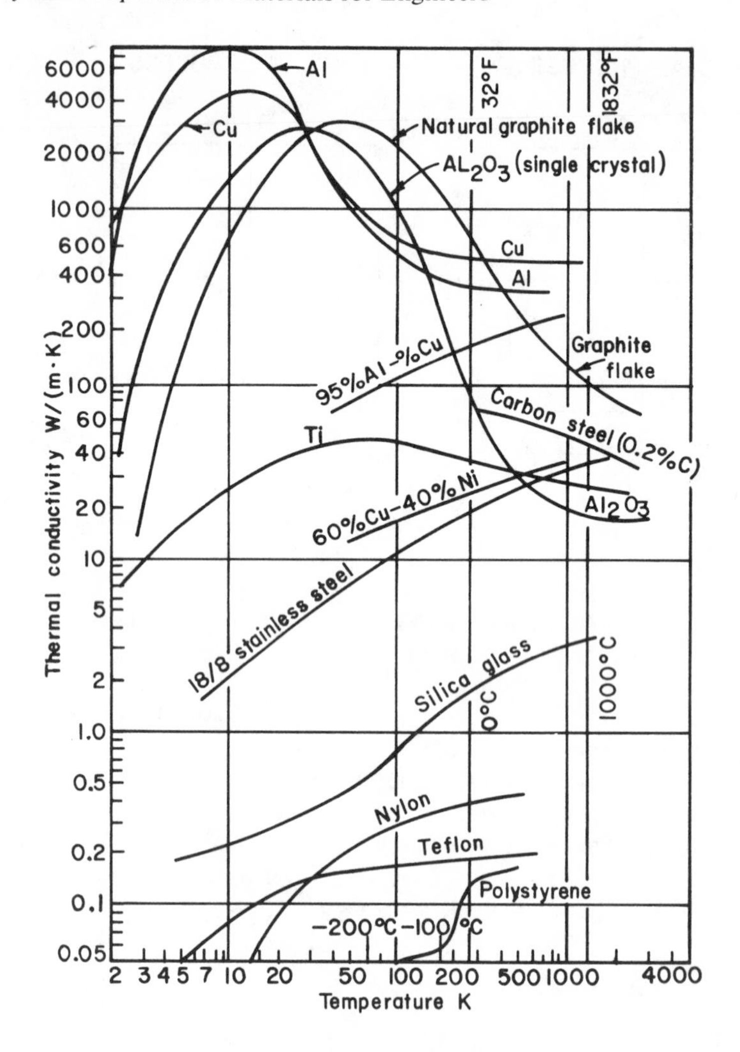

FIGURE 4-23. Thermal conductivities of some typical materials. (After Jastrzebski, Z. D., *The Nature and Properties of Engineering Materials,* John Wiley & Sons, New York, 1976, 504. With permission.)

the resultant vector in a normal interaction is not affected by imperfections or crystal boundaries, such as previously discussed for very low temperatures, the thermal conductivity would be expected to be relatively high.

Umklapp processes occur when the resultant of the two interacting phonons lies outside of the first Brillouin zone. The zone vector is added to the resultant in order to bring it back into the first zone. Large momentum changes occur when phonons with large wave vectors interact; the flux decreases and a decrease occurs in the effectiveness in the transmission of the energy through the solid. This would be expected to become increasingly inefficient as the number of umklapp interactions increases. By the same token, where momentum is unchanged, as in normal processes, the energy is more effectively transmitted. The picture provided by the Brillouin zone model of a solid provides a simple means for understanding the behavior of phonons in solids.

4.4.1. Commercial Materials

The very large number of thermal insulating materials used in engineering applications makes it impossible to provide detailed data here; these are available in hand-

Table 4-4
THERMAL CONDUCTIVITIES OF SOME SELECTED CERAMIC AND ORGANIC MATERIALS[a]

Material	Thermal conductivity (cal/sec) (cm²) (°C/cm) 37.8°C (100°F)	93.3°C (200°F)	148.9°C (300°F)
Single crystals			
Silicon Carbide	0.21	0.21	0.20
Periclase	0.11	0.09	0.08
Spinel $MgO \cdot Al_2O_3$	0.03	0.03	0.02
Quartz (c axis)	0.03	0.02	0.02
Quartz (basal plane)	0.01	0.01	0.01
Fluorite	0.02	0.02	0.01
Polycrystalline materials			
BeO (pure, hot pressed)	0.52	0.43	0.38
MgO (spec. pure)	0.09	0.08	0.07
ThO_2 (hot pressed)	0.04	0.03	0.03
PbO	0.007	0.005	0.004
Organic materials[b]		K (near room temperature)	
Polyethylene		0.002	
Rubber		0.007	
Urethane foam		0.0001	

[a] Abstracted from Reference 8.
[b] Abstracted from Reference 9.

books. The thermal conductivities of some typical materials are shown in Figure 4-23. The high conductivities of aluminum and copper are to be expected because of the transport properties of their nearly free valence electrons. The behavior of graphite is explained by the trigonal s-p hybridized bonding. The weak bond between the basal planes is readily available for conduction (see Sections 3.11.2, Chapter 3, 8.2.1.2, Chapter 8 and 10.6.7, Chapter 10). The comparably high properties of the single Al_2O_3 crystal results from phonon conduction due to its unusual purity and crystalline perfection.

The effects of impurities, in the materials shown in the figure, are the result of intentionally added alloying elements. These are shown by the properties of 95% Al-5% Cu and 60% Cu-40% Ni when compared to those of the respective pure elements. The comparison of the properties of the 18-8 stainless steel with those of the 0.2C steel also shows a significant contrast which results from large differences in alloying elements.

The organic polymeric materials show the lowest thermal conductivities because of the strong covalent bonding involved in these materials; the energy transfer is almost entirely by phonons in these virtually noncrystalline solids.

Data for some typical ceramic and organic materials are given in Table 4-4 (see also Table 4-3).

4.5. PROBLEMS

1. Is Equation 4-7 valid for gases? Give reasons.

2. Derive a relationship between C_p and C_V for gases.
3. How would C_V be expected to change for a solid going through (a) an allotropic change, (b) ordering, (c) a magnetic transition?
4. What is the meaning of the gas constant R?
5. Show that Equation 4-41 is a solution of Equation 4-39.
6. Contrast the classical, Einstein, Born-von Kármán, and Debye models for the internal energies of solids. What is the significance of the differences between them?
7. Explain the basis for the failure of the Einstein model of the heat capacity of solids at low temperatures.
8. How many normal modes can be expected for a particle that has both rotational and translational freedom?
9. Use a sketch to demonstrate the Brillouin zone concept for velocities external to the first zone as given in Figure 4-6.
10. Discuss the limitations of the Born-von Kármán model for the internal energy of a solid.
11. Use Equations 4-79a and 4-79b to obtain expressions for the evaluation of C_V at 0 K and at relatively high temperatures, using the ideas given in the text.
12. What suggestions could be made for the application of Equation 4-79 to a polycrystalline solid?
13. Under what conditions might the approximation given by Equation 4-92 be made with confidence?
14. Obtain an approximation for the minimum Debye wavelength for an average metallic lattice.
15. Explain the partial success of the classical theory for heat capacity.
16. Explain how the grain size of a polycrystalline material might influence experimentally obtained heat capacity data at very low temperatures.
17. Explain the basic difference in the ways classical and quantum mechanics employ to explain thermal expansion.
18. Explain the effects of the approximations made for coefficient of expansion of a solid on the resultant expression.
19. Express the Grüneisen constant as a function of frequency.
20. Explain the conditions under which a nonmetallic solid would be a better conductor of heat than a metal. How might these ideas be applied to cryogenic insulation systems?
21. Discuss the ideas implicit in the assumption made to obtain Equation 4-145.
22. Explain the ways in which C_V, v, and λ are interrelated.
23. Compare the behaviors of nonconducting solids with metals at elevated temperatures and explain the differences.

4.6. REFERENCES

1. **Sproull, R. L.,** *Modern Physics,* John Wiley & Sons, New York, 1956.
2. **Richtmyer, F. K., Kennard, E. H., Lauritsen, T.,** *Introduction to Modern Physics,* 5th ed., McGraw-Hill, New York, 1955.
3. **Dekker, A. J.,** *Solid State Physics,* Prentice-Hall, Englewood Cliffs, N. J., 1959.
4. **Kittel, C.,** *Introduction to Solid State Physics,* John Wiley & Sons, New York, 1966.
5. **Darken, L. S. and Gurry, R. W.,** *Physical Chemistry of Metals,* McGraw-Hill, New York, 1953.
6. **Sokolnikoff, I. S. and Redheffer, R. M.,** *Mathematics of Physics and Modern Engineering,* McGraw-Hill, New York, 1958.

7. **Butts, A.,** *Metallurgical Problems,* McGraw-Hill, New York, 1943.
8. **Weast, R. C., Ed.,** *Handbook of Chemistry and Physics,* 56th ed., CRC Press, Cleveland, Ohio, 1975.
9. **Guy, A. G.,** *Essentials of Materials Science,* McGraw-Hill, New York, 1976.
10. **de Launay, J.,** *Solid State Physics,* Vol. 2, Seitz, F. and Turnbull, D., Eds., Academic Press, New York, 1956.
11. **Kelley, K. K.,** Bureau of Mines Bulletin No. 584, U.S. Government Printing Office, Washington, D.C., 1960.

Chapter 5

CLASSIFICATION OF SOLIDS

The historical approach will be taken to describe the principle classes of solids in terms of the electron mechanisms responsible for their physical properties. This will permit insight into the origin and growth of the modern fundamental concepts. The classical theory will be used as a point of departure to develop and examine the modern theory. This will afford a better understanding of the properties of solids and permit their classification based upon the configuration and behavior of their valence electrons.

Seitz's classification of solids, based upon physical and mechanical properties, gives a brief overview. Metals have positive temperature coefficients of electrical resistivity, high electronic, thermal, and electrical conductivities, and can be deformed plastically. Ionic crystals are electrically neutral assemblies of ions from both ends of the periodic table, i.e., LiF and NaCl. These have low ionic rather than electronic electrical and thermal conductivities and cleave rather than deform. Covalent (valence) crystals are usually hard and abrasive, cleave readily and have low electrical and thermal conductivities. Solids such as diamond and silicon carbide are in this class. Semiconductors (Ge, Si, etc.) also are covalently bonded and have low electrical conductivities with negative temperature coefficients. Other solids are bound by molecular, or van der Waals, forces. The electron behavior responsible for the properties of metals, semiconductors, and insulators is discussed here. Some properties of ionic and covalent crystals and compounds are presented in Chapters 8 (Volume II) and 10 (Volume III). Selected properties of semiconductors and insulators are discussed in Chapters 4, and Chapters 11 and 12, Volume III.

5.1. DRUDE-LORENTZ THEORY OF METALS

As early as 1900 Drude suggested that the high electrical conductivity of metals could be explained in terms of their valence electrons. These were considered to be free to move within the solid in a way similar to gas molecules in a container. This idea is still correct in a qualitative way. Drude's ideas were developed further by Lorentz.

The Drude-Lorentz theory is based upon a model similar to that used to explain the photoelectric effect (Figure 1-2 in Chapter 1). The electrons were free to roam about within the metal crystal lattice in a constant, internal potential. Potential barriers were considered to exist at the metallic surfaces to retain the electrons, since the electrons do not ordinarily leave the metal. The energy levels of the electrons within the metal must, therefore, be lower than those of the energy barriers at the surfaces. This difference in energy is the work function, as discussed in Chapter 1. It is the minimum energy which must be imparted to an electron in order to remove it from the metal. This is one of the bases for the models which treat electrons as being in a "box", or in a potential well.

All free valence electrons within the metal were considered to behave in the same way as gas molecules, i.e., to obey the classical Maxwell-Boltzmann statistics. As previously noted, this assumed that the electrons had a continuous energy distribution.

The flow of such electrons was considered to be impeded only by collisions with the ions of the metal; electron-electron interactions were not included. The main transfer of energy within the solid was considered to arise from electron-ion collisions. The energy involved was thought to be small since the mass of an electron is so very much smaller than that of an ion core. At thermal equilibrium, an equilibrium was also considered to be reached in the exchange of such energy. The average thermal velocity of an electron considered in this way could be established from the Maxwell-Boltzmann statistics.

5.1.1. Electrical Resistivity (Conductivity)

The application of an electric field (voltage/length) to a metallic conductor was considered to cause *all* of the free valence electrons (carriers) to flow in the metal and thereby constitute a current. The presence of this field, $\bar{E}_x$, accelerates the electrons and provides an extra component of velocity parallel to the field, or drift velocity, V, in addition to their average thermal velocity, μ. This thermal velocity is considered to be unaffected by the electric field. An accelerated electron collides with an ion in the lattice and energy is absorbed as a phonon. This energy absorption is considered to take place such that the electron is assumed to have a zero velocity component parallel to the electric field immediately after the collision. The electron then regains its drift velocity in addition to its thermal velocity and proceeds through the lattice in a random direction until another collision takes place. This process is repeated by all of the free valence electrons. The result is a net flow of electrons in the direction opposite to that of the field. This electron flow constitutes the electric current.

The acceleration of the electrons which results from the electric field is

$$a_x = \frac{\partial V}{\partial t} = \frac{F}{m} = \frac{\bar{E}_x e}{m} \tag{5-1}$$

The force, F, is given by the product of the electric field, $\bar{E}_x$, and the charge, e, on the electron; m is the mass of the electron. The drift velocity is obtained by integrating Equation 5-1.

$$V = \frac{\bar{E}_x e}{m} t \tag{5-2}$$

Since the drift velocity varies from zero just after collision to V at an instant prior to the next collision, the average drift velocity may be approximated by

$$\bar{V} = \frac{V}{2} = \frac{\bar{E}_x e}{2m} t \tag{5-2a}$$

If the electron travels an average (mean free path) L between collisions, then the average time between collisions may be obtained from

$$t = \frac{L}{\mu + \bar{V}}$$

However, $\bar{V}$ is much smaller than μ so that it may be approximated that

$$t \simeq \frac{L}{\mu} \tag{5-3}$$

Equation 5-2 may be expressed, using Equation 5-3, as

$$\bar{V} = \frac{\bar{E}_x e}{2m} \cdot \frac{L}{\mu} \tag{5-4}$$

The expression for the current density is

$$j = ne\bar{V} \tag{5-5}$$

where n is the number of free valence electrons in a unit volume of the metal. Equation 5-4 is substituted into Equation 5-5 to obtain

$$j = ne\frac{\bar{E}_x eL}{2m\mu} = \frac{ne^2 L\bar{E}_x}{2m\mu}$$

This relationship is compared to Ohm's law to arrive at an expression for the electrical resistivity:

$$j = \frac{\bar{E}_x}{\rho} = \frac{ne^2 L\bar{E}_x}{2m\mu}$$

The electric field, $\bar{E}_x$, vanishes and

$$\rho = \frac{2m\mu}{ne^2 L} = \frac{1}{\sigma} \tag{5-6}$$

The temperature dependence of ϱ should be calculated from the average velocity of the electrons as determined from a Maxwell-Boltzmann distribution; this is consistent with the behavior of the electrons as postulated in the previous section. However, the approximations made to obtain Equations 5-2a and 5-3 are reflected in Equation 5-6, so that a high degree of rigor is unwarranted. A less complicated approach may be used to obtain an expression with the same temperature dependence as is provided by the more rigorous method. This is accomplished by means of energy relationships from the ideal-gas theory:

$$E = m\mu^2/2 = 3k_BT/2$$

The temperature dependence of μ, found from these equations, is

$$\mu = \left[\frac{3k_BT}{m}\right]^{1/2} \tag{5-7}$$

The equation for the electrical resistivity of a metal, obtained by the substitution of Equation 5-7 into Equation 5-6 is found to be

$$\rho = \frac{2m}{ne^2 L}\left[\frac{3k_BT}{m}\right]^{1/2} = \frac{2}{ne^2 L}(3mk_BT)^{1/2} = \frac{1}{\sigma} \tag{5-8}$$

It should be noted that the more rigorous approach gives the coefficient as $3\pi/8$ instead of 2 in Equation 5-8.

On the basis of Equation 5-8 it is expected that the electrical resistivity of a metal should be a function of $T^{1/2}$. This prediction is erroneous. Most pure metals have electrical resistivities which are very close to being linear functions of temperature at all temperatures higher than about $0.2\theta_D$. In fact, most of these have temperature coefficients of electrical resistivity of about 0.4% per degree in this range of temperatures (see Section 6.2 and Table 6.1 in Chapter 6, Volume II).

Despite the fact that Equation 5-8 gives a wrong temperature dependence, calculations based upon it give values for ϱ which are in approximate agreement with the experimental data taken near room temperature. This agreement, as is shown by examination of Sections 6.1 and 6.2 in Chapter 6 is purely one of chance. However, Equation 5-8 is of importance because it is one of the factors which demonstrate that the Drude-Lorentz theory contains an inherent error (see Section 5.1.3). As such, it gave impetus to the search for models of electron behavior which are more accurately descriptive of the properties of metals.

5.1.2. Thermal Conductivity

The valence electrons, being free, should contribute to the thermal conductivity of a metal as well as the ions (Section 4.4 in Chapter 4). The application of the ideal-gas laws to the free valence electrons gives the kinetic energy of one electron as

$$E = \frac{3}{2} k_B T$$

The derivative, taken with respect to the direction of heat flow in the x direction, is

$$\frac{\partial E}{\partial x} = \frac{3}{2} k_B \frac{\partial T}{\partial x} \tag{5-9}$$

And, where L is the mean free path of an electron, the transmission of energy per electron is

$$\frac{\partial E}{\partial x} L = \frac{3}{2} k_B \frac{\partial T}{\partial x} L \tag{5-10}$$

The velocity of a single electron, treated as a gas molecule, in terms of its component velocities, is

$$\mu^2 = \mu_x{}^2 + \mu_y{}^2 + \mu_z{}^2$$

However, the large number of valence electrons travel in random paths in the metal between elastic collisions with the ions. Their motion in any one of the cartesian directions is as equally probable as in either of the other two directions. Under these conditions, the average velocity of an electron moving in the x direction is $\mu_x = 1/3\ \mu$. This gives the flux, or the number of electrons crossing a unit area in unit time, as $n\mu/3$.

Let the total thermal energy transmitted across a unit area (perpendicular to the x direction) in unit time be given by Q. Then

$$Q = \text{flux} \times \text{energy transmitted/electron}$$

Substitutions are made for the flux and for the transmission of energy, Equation 5-10, to give

$$Q = \frac{n\mu}{3} \cdot \frac{3}{2} k_B \frac{\partial T}{\partial x} L = \frac{1}{2} n\mu\, k_B \frac{\partial T}{\partial x} L \tag{5-11}$$

Equation 4-148 is used in the form

$$\kappa = Q \frac{\partial x}{\partial T} = \frac{Q}{\frac{\partial T}{\partial x}} \quad (5\text{-}12)$$

Equation 5-11 is substituted into Equation 5-12 to obtain thermal conductivity as

$$\kappa = \frac{\frac{1}{2} n\mu\, k_B \frac{\partial T}{\partial x} L}{\frac{\partial T}{\partial x}} = \frac{1}{2} n\mu\, k_B\, L \quad (5\text{-}13)$$

Here the variables are μ and L. These are functions of temperature and provide the temperature dependence of κ

The Wiedemann-Franz law states that the ratio of the thermal conductivity to the electrical conductivity is a "constant" (see Section 5.6.5). So, if Equations 5-6 and 5-13 are correct, their ratio should predict the desired constant. This ratio is

$$\frac{\kappa}{\sigma} = \frac{\frac{1}{2} n\mu\, k_B L}{\frac{ne^2 L}{2m\mu}} = \frac{k_B m\mu^2}{e^2}$$

Equation 5-7 is used to obtain an expression for μ^2 which results in

$$\frac{\kappa}{\sigma} = \frac{k_B m}{e^2} \cdot \frac{3k_B T}{m} = 3\left[\frac{k_B}{e}\right]^2 T$$

or,

$$\frac{\kappa}{\sigma T} = 3\left[\frac{k_B}{e}\right]^2 = L_o \quad (5\text{-}14)$$

L_o is the constant required by the Wiedemann-Franz law and is known as the Lorentz number. The Lorentz number may be expressed as 2.45×10^{-8} watt-ohm/deg^2, 0.57×10^{-8} cal-ohm/sec-deg^2 or as 2.73×10^{-13} esu/deg^2 (see Table 5-1). Equation 5-14 provides a good approximation of this. It will be seen later that this result constitutes the major accomplishment of the Drude-Lorentz theory.

5.1.3. Heat Capacity

The contribution of electrons to the internal energy of a solid, U_e, must be added to that of the ions (Equation 4-9) to obtain the total energy of that solid, i.e.,

$$U = U_e + U_i$$

Still treating the free electrons as an ideal gas, this becomes

$$U = 3/2\, N' k_B T + 3N k_B T \quad (5\text{-}15)$$

where the number of valence electrons is

$$N' = ZN \quad (5\text{-}15a)$$

Table 5-1
EXPERIMENTAL LORENTZ NUMBERS

	$L_o \times 10^8$ w-Ω/deg²			$L_o \times 10^8$ w-Ω/deg²	
Metal	0°C	100°C	Metal	0°C	100°C
Ag	2.31	2.37	Pb	2.47	2.56
Au	2.35	2.40	Pt	2.51	2.60
Cd	2.42	2.43	Sn	2.52	2.49
Cu	2.23	2.33	W	3.04	3.20
Ir	2.49	2.49	Zn	2.31	2.33
Mo	2.61	2.79			

From Kittel, C., *Introduction to Solid State Physics*, 5th ed., John Wiley & Sons, New York, 1976, 222. With permission.

in which Z is the valence of the metallic atom and N is Avogadro's number. Then the heat capacity is

$$C_V = \left[\frac{\partial U}{\partial T}\right]_V = 3/2\ N'k_B + 3Nk_B \qquad (5\text{-}16)$$

If a univalent metal is being considered, then N′ = N and

$$C_V = 3/2\ Nk_B + 3Nk_B = 9/2\ Nk_B = 9/2\ R \simeq 9 \text{ cal/mol/deg}$$

It was previously shown (Chapter 4) that $C_V \simeq 6$ cal/mol/degree. Therefore, Equation 5-16 is in error by at least 50%. This error would increase according to the number of valence electrons associated with the metallic atom.

This poses a dilemma. The number of electrons calculated in this way gives approximate agreement of resistivity (Equation 5-8) with experiment at room temperature. The same type of calculation for heat capacity (Equation 5-16) gives a minimum error of 50%. It could be argued that N′ is very small in order to arrive at a correct value for C_V. But, when this is done, the calculation for resistivity is too small. The theory gives inconsistent results. Evidently, the Drude-Lorentz theory contains a fundamental error. The root of this problem, and its elimination, is discussed in the section on the Fermi-Sommerfeld theory.

5.1.4. Summary

The Drude-Lorentz theory is based upon the idea that the behavior of electrons in a solid is comparable to that of molecules of ideal gases. It results in an incorrect temperature dependence of resistivity ($\varrho \propto T^{1/2}$), provides a reasonably accurate representation of the Wiedemann-Franz ratio, and gives a value for the heat capacity of a metal which is much too great. The calculations for the electrical resistivity and heat capacity reveal that this approach contains a fundamental inconsistency. In addition, this theory for the properties of metals is not capable of describing other types of solids.

5.2. SOMMERFELD THEORY OF METALS

The Sommerfeld theory marks the real beginning of the application of quantum mechanics as a means for understanding the physical properties of solids. Metals are treated as continuous, homogeneous, isotropic solids. The electron picture used is similar to that employed by the Drude-Lorentz model; the free electrons are considered to be in a well of constant internal potential. The significant difference is that the electrons retained in the well obey quantum, instead of classical, mechanics. This gives further insight into the reasons for the boundary conditions imposed on the solutions to Schrödinger's equation (Chapter 3).

It will be recalled that the presence of a large number of electrons in a potential well whose dimensions are of crystal sizes would result in a quasi-continuum of energy levels. The Pauli Exclusion Principle dictates the way in which these energy states are filled. Neglecting spin, only two electrons can occupy each state. If a crystal of an element of the order of 1 mol wt is being considered, about 6×10^{23} electrons must be taken into account. Thus, starting with the lowest energy level, about 3×10^{23} states must be filled consecutively until all electrons occupy states. This means that only a small fraction of the total number of electrons can occupy energy states in the neighborhood of those with work functions close to W in Figure 1-2, Chapter 1.

The quantum-mechanic rules which state that electrons can change their energy only by discrete amounts hold here. Thus, those electrons with work functions considerably greater than W cannot enter into a physical process unless abnormally large amounts of energy are supplied. Vacant states are not available for the occupation of such low-energy electrons when normal amounts of energy are applied. Only those electrons occupying states with work functions close to W are available for participation in physical processes, when energies of the usually encountered magnitudes are applied, because vacant states just above these are available for them to occupy. Recall that the Drude-Lorentz theory included all the free electrons.

The question then becomes one of the determination of the ways in which the electrons enter into physical processes such as thermal and electrical conductivity, etc. This must be resolved into two other questions:

1. How many states are available for electron occupation? Or, in other words, what is the density of states?
2. How are the electrons distributed in these states and how do they enter into physical processes?

Attention will now be directed to the first of these questions, the density of electron states, by beginning with Equation 3-49 for a three-dimensional well

$$E = \frac{h^2}{8m}\left[\frac{n_x^2}{L_x^2} + \frac{n_y^2}{L_y^2} + \frac{n_z^2}{L_z^2}\right]$$

$$= \frac{1}{2m}\left[\left(\frac{n_x h}{2L_x}\right)^2 + \left(\frac{n_y h}{2L_y}\right)^2 + \left(\frac{n_z h}{2L_z}\right)^2\right] \qquad (3\text{-}49)$$

where each state implicitly includes two spins. The energy also may be expressed in terms of momentum as

$$E = \frac{p^2}{2m} = \frac{p_x^2 + p_y^2 + p_z^2}{2m} \qquad (5\text{-}17)$$

Equations 3-49 and 5-17 are equated to give

$$p_x^2 + p_y^2 + p_z^2 = \left[\frac{n_x h}{2Lx}\right]^2 + \left[\frac{n_y h}{2L_y}\right]^2 + \left[\frac{n_z h}{2L_z}\right]^2$$

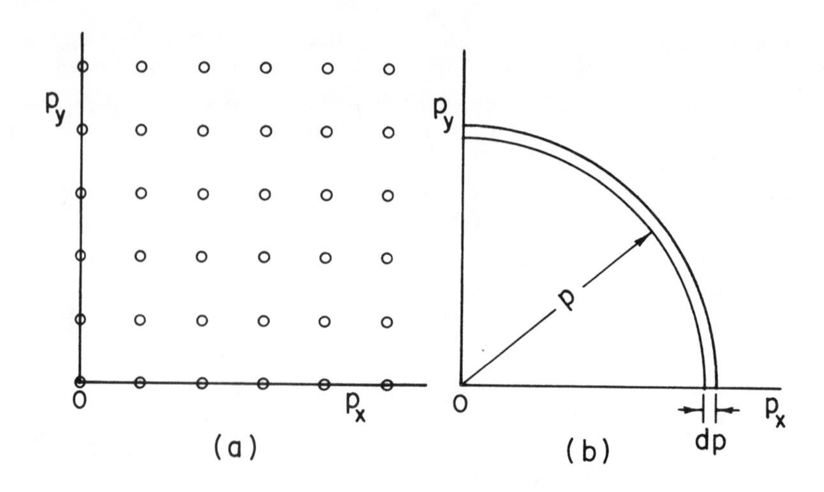

FIGURE 5-1. (a) Two-dimensional momentum space magnified to show quantized states; (b) basis for calculating N(p).

so that

$$p_i = \pm \frac{n_i h}{2L_i} \tag{5-18}$$

The ± sign refers to the fact that the electron can be moving in either the positive or negative directions. Equation 5-18 quantizes momentum space. Now if the well is considered to be a cube of atomic dimensions, the unit of momentum space is h/2L. Consider the first quadrant of momentum space, in two dimensions (Figure 5-1a). Here each point represents an allowed momentum state. Each momentum state occupies an area of $h^2/4L^2$ in the figure. Three-dimensional momentum space must be employed in order to satisfy Equation 5-17. Here the volume occupied by each state is $h^3/8L^3$. Only the first octant is considered because the other octants give redundant values of momentum. Thus, the function for the density of momentum states can be determined from the volume between p and p + dp and the volume of each state expressed by

$$N(p)\,dp = \frac{\text{volume between p and p + dp}}{\text{volume of one momentum state}} = \frac{\dfrac{4\pi p^2 dp}{8}}{\dfrac{h^3}{8L^3}}$$

which reduces to the number of states between p and p + dp given by

$$N(p)\,dp = \frac{4\pi V p^2 dp}{h^3} \tag{5-19}$$

since $L^3 = V$, the volume of one well. It is more convenient to describe the density of states in terms of energy rather than in terms of momentum. This conversion may be accomplished from the classical relationships

$$p = [2mE]^{1/2} \text{ and } dp = m[2mE]^{-1/2}\,dE \tag{5-20}$$

These are substituted into Equation 5-19 to give

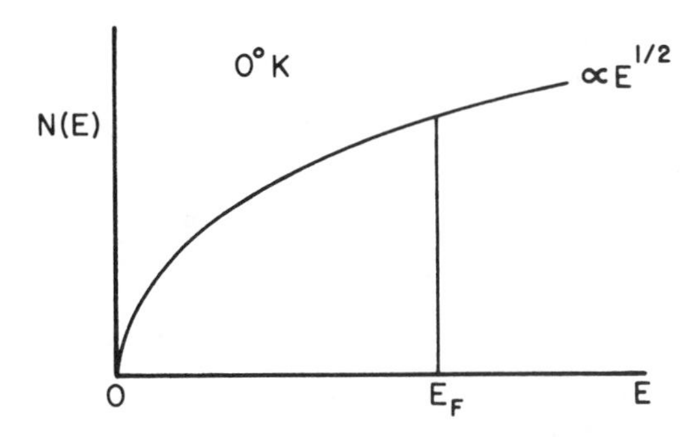

FIGURE 5-2. Density of states of electrons at 0 K.

$$N(E)dE = \frac{4\pi V}{h^3} \cdot 2mE \cdot m\,[2mE]^{-1/2}\, dE$$

Upon rearranging, this becomes

$$N(E)dE = \frac{2\pi V}{h^3}(2m)^2\,(2m)^{-1/2}E^{1/2}\, dE$$

or

$$N(E)dE = \frac{2\pi V}{h^3}(2m)^{3/2}\, E^{1/2}\, dE = dN \tag{5-21}$$

Here m actually is the "effective mass" of the electron; it can be significantly different from that of a free electron. This property, usually designated by m*, is discussed more fully later in this chapter, in the section on Brillouin zones, and in Chapter 11 in Volume III.

Basically, Equation 5-21 is the answer to the question regarding the number of states available for occupation by electrons; this equation is the Sommerfeld distribution function, where each state includes two spins.

5.3. THE FERMI LEVEL

It will be recalled (Figure 3-4 in Chapter 3) that when the crystal size becomes large, the energy levels form a quasi-continuum. In addition, according to the Pauli Exclusion Principle, only two electrons can occupy each level if spin is implicitly included. Under these conditions the right-hand side of Equation 5-21 need not be multiplied by a factor of two, since two spins were originally included in this application of Equation 3-49. The curve for the density of states given by Equation 5-21 is shown schematically in Figure 5-2.

As previously noted, the energy states are filled with valence electrons, starting with the lowest state, until all of the electrons are accommodated. The electron in the highest energy state to be filled at 0 K is analogous to the electron at the highest energy level in the well shown in Figure 1-2 in Chapter 1. Such an electron would require the energy increment W in order to leave the well. This energy state, i.e., the highest level to be occupied by an electron at 0 K, is tentatively defined as the Fermi level, E_F. As will be seen later, this "definition" is incomplete. It is, nevertheless, sufficient for present purposes (see Section 5.5).

The spherical model, illustrated in two dimensions in Figure 5-lb, can be used to determine the number of occupied energy states in a somewhat different way. Here, the radius vector, p, is the distance from the origin to the outermost filled state. The unit of momentum space is obtained from the de Broglie relationship, $p = h/\lambda$, now as given by $p = h/L$. Using the spherical model, the number of occupied states is obtained by calculating the number of momentum states in the volume occupied in momentum space:

$$N = 2 \cdot \frac{4}{3}\pi p^3 \div \frac{h^3}{L^3} = \frac{8\pi p^3 V}{3h^3} \tag{5-22}$$

Here the factor 2 is used explicitly to include two spins in each state. The relationship given by Equation 5-20 is used again to obtain

$$N = \frac{8\pi (2m)^{3/2} E^{3/2} V}{3h^3} \tag{5-23}$$

The differentiation of Equation 5-23 with respect to E gives Equation 5-21 when spin is taken into account. Conversely, Equation 5- 21 may be integrated to give Equation 5-23 for the conditions noted.

Either of the above equations for N can be solved to obtain an expression for E_F. Rewriting Equation 5-23

$$E_F^{3/2} = \frac{1}{8} \cdot \frac{3N}{\pi V} \left[\frac{1}{2m}\right]^{3/2} h^3$$

then

$$E_F = \frac{1}{4} \left[\frac{3N}{\pi V}\right]^{2/3} \left[\frac{1}{2m}\right] h^2 = \frac{h^2}{8m} \left[\frac{3N}{\pi V}\right]^{2/3} \tag{5-24}$$

If the factor V is taken as the volume of an ion in a crystal lattice, then N/V is the electron:ion ratio. Thus, E_F may be considered to be a function of the electron:ion ratio. This concept will be useful in the discussion of the role of the Fermi level on the physical properties of alloys and on the formation of alloy phases.

It is interesting and instructive to obtain an approximation of the average energy of an electron. This can be done by means of classical statistics, where the average energy is given by

$$\bar{E} = \frac{\int_0^{E_F} EN(E)dE}{\int_0^{E_F} N(E)dE} = \frac{\int_0^{E_F} E^{3/2}dE}{\int_0^{E_F} E^{1/2}dE}$$

since the constant factors in Equation 5-21 vanish. Thus, the average energy of the valence electrons is

$$\bar{E} = \frac{\frac{2}{5} E_F^{5/2}}{\frac{2}{3} E_F^{3/2}} = \frac{3}{5} E_F \tag{5-25}$$

Since only those electrons with $E \geqslant E_F$ can enter into physical processes, it is of

Table 5-2
CALCULATED FERMI ENERGIES AND TEMPERATURES FOR FREE ELECTRONS[2]

Element	E_F (eV)	T_F (°K × 10^{-4})
Li	4.7	5.5
Na	3.1	3.7
K	2.1	2.4
Rb	1.8	2.1
Cs	1.5	1.8
Cu	7.0	8.2
Ag	5.5	6.4
Au	5.5	6.4

Note: leV = 23,050 cal/mol.

interest to approximate the value of E_F. Such an approximation can be obtained readily from Equation 5-24 for a normal, monovalent metal with a typical atomic diameter of about 3 Å. Using $h \simeq 6.6 \times 10^{-27}$ erg sec, $m \simeq 9.1 \times 10^{-28}$ g, and $V \simeq (3 \times 10^{-8})^3$ cm³ results in $E_F \simeq 6.5 \times 10^{-12}$ erg or, using 1 eV $\simeq 1.6 \times 10^{-12}$ erg, $E_F \simeq 4$ eV.

The magnitude of E_F may be more readily appreciated by considering the properties of an electron with an energy equal to E_F in terms of those of a classical electron. This is accomplished by means of the equation for the energy of a classical electron in the form $E_F = k_B T_F$, where T_F is the Fermi temperature. Using $E_F \simeq 4$ eV and $k_B \simeq 8.6 \times 10^{-5}$ eV/K gives $T_F \simeq 4.7 \times 10^4$K. In essence, this means that a classical electron would have to be at about 50,000K to possess the energy equivalent of an electron at E_F. The rough approximations of E_F and T_F given here lie close to the averages of the more accurately calculated values given in Table 5-2.

It is seen that the average energy of an electron is quite high. The addition of normal energies of the order of k_BT will have only a small effect upon the electrons, and then only upon those with the highest energies. The effect of temperature upon the Fermi level is given by

$$E_F(T) = E_F(0)\left[1 - \frac{\pi^2}{12}\left(\frac{k_BT}{E_F(0)}\right)^2\right] \qquad (5\text{-}26)$$

The factor k_BT is very small with respect to $E_F(0)$ for most metals, so that the fraction within the brackets is very small. Thus, E_F decreases very slightly with increasing temperature. For most cases, this decrease is so small that it may be neglected, and E_F usually is considered to be virtually invariant with temperature. One exception to this is discussed in Section 7.9.4, Chapter 7, Volume II.

5.4. STATISTICAL APPROACH

The question regarding the density of states available for occupation by electrons has been answered by Equation 5-21. The second question, the way in which the electrons are distributed in these states, has been answered for 0 K, but not for other temperatures. In addition, the ways in which the electrons are considered to participate in physical process have not been considered yet.

The behavior of the electrons at 0 K can be described with the help of Figure 5-2. For all energies less than E_F, the density of states at 0 K, N(E,0), is just twice N(E) (Equation 5-21) when spin is explicitly considered. In addition, for all energies greater than E_F, N(E,0) must equal zero. This must be the case, since E_F was found by consecutively filling all states, starting with the lowest, until *all* of the electrons were taken into account. Therefore, all states above E_F must be unoccupied. In other words, at 0 K,

$$\text{for } E < E_F$$

$$N(E,0) = 2N(E) \tag{5-27a}$$

$$\text{and } E > E_F$$

$$N(E,0) = 0 \tag{5-27b}$$

It will be noted that Equation 5-21 only provides a parabolic relationship for the density of an indefinite number of states. It must be modified to include the behaviors noted by Equations 5-27a and 5-27b. A partition, or cut-off, function can be defined to include this behavior. Such a function, f(E,0) can be defined such that

$$E < E_F$$
$$N(E,0) = 2N(E,0) \cdot f(E,0) = 2N(E) \tag{5-28a}$$

and

$$E > E_F$$
$$N(E,0) = 2N(E,0) \cdot f(E,0) = 0 \tag{5-28b}$$

Thus, f(E,0) must be of the form

$$f(E,0) = f(E,T) = \left| \begin{array}{l} 1 \text{ for } E < E_F \\ \\ 0 \text{ for } E > E_F \end{array} \right. \tag{5-29}$$

Here f(E,T) includes the effect of temperatures beyond the absolute zero.

Since this involves the energies of electrons in potential wells, it will be recalled that the Heisenberg Uncertainty Principle, as well as the Pauli Exclusion Principle, must be taken into account in determining f(E,T). How is such a function found? It turns out that the essence of the problem is to decide upon the number of ways that indistinguishable particles (electrons) can be placed *singly,* because of the Exclusion Principle, into distinguishable "boxes" (energy states). What statistical approach shall be used here? In order to answer this question the capabilities of classical statistics will be examined first and compared to the conditions which must be imposed upon quantum statistics. Factors such as indistinguishability and exclusion will, of necessity, be taken into account. The Fermi-Dirac statistics then will be developed.

5.4.1. Classical Statistics

The Maxwell-Boltzmann, or classical, statistics are based upon the assumption that each particle in an assembly can be counted or identified. There is no uncertainty about finding the particle or regarding its energy. No restrictions are placed upon the energies of the particles; these are expected to represent a continuous distribution of energy.

Thus, due to the absence of any limitations, no conditions are placed upon the degree to which the momenta, position, or energies of such particles can be known. (The state space describing these particles is six-dimensional: x, y, z, p_x, p_y, p_z.) Under these conditions, the particles may occupy *any* position in state space. Multiple occupancy of a "cell" in state space is permitted. This means that more than one particle can have the same coordinates in state space and thus occupy the same state.

Examples of where this statistical approach may be used include the elastic vibrations of ions in solids and molecules of an ideal gas. This model can give representative results when distinguishability is not considered, any actual energy discontinuities are small with respect to k_BT, and there is small probability of multiple occupancy of a given state. The latter condition may also be expressed in terms of the number of particles per unit of state space being small. This, for example, may be taken as a small number of particles per energy or momentum range.

An illustration of this may be based upon the Einstein model. It was shown that

$$\bar{E} = \frac{h\nu_i}{\exp(h\nu_i/k_BT) - 1} \tag{4-19}$$

From this, the probable number of particles with energy $h\nu_i$ (the probable occupancy of a given state), is

$$N_i = \frac{\bar{E}}{h\nu_i} = \frac{1}{\exp(h\nu_i/k_BT) - 1} \approx e^{-h\nu_i/k_BT} = e^{-E_i/k_BT}$$

It can be reduced to the Boltzmann factor, as shown, when the exponential term is much greater than unity. Conditions can be arranged such that the probability of multiple occupancy of a state is very low. In order that the probability of single occupancy of a state be high, the condition

$$N = \frac{1}{V}$$

must be met, where N is the number of particles per unit volume of state space and V is the maximum volume of state space per particle. It should be noted that this is *not* equal to the volume of the particle. In addition, the mean distance between the particles, λ, should be such that

$$\lambda \ll \frac{1}{N} \tag{5-30}$$

then

$$\lambda \ll V^{1/3} \text{ and } \lambda^3 \ll V \tag{5-31}$$

Under these restrictions, the probable occupancy of state E_i becomes

$$N_i = N\lambda^3 e^{-E_i/k_BT}$$

Now, substituting Equations 5-30 and 5-31 into this gives

$$N_i = \frac{\lambda^3}{V} e^{-E_i/k_BT} \tag{5-32}$$

If $\lambda^3 = V_i$ represents the minimum volume of state space which can be assigned to each particle, the probable occupancy of a state is given by

$$N_i = \frac{V_i}{V} e^{-E_i/k_B T} \tag{5-33}$$

This kind of control upon λ or V_i permits the adjustment of state space in such a way that each particle can have a high probability of occupying a single state. In effect, since the states can be made to be very small, and the distances between the centers of adjacent states equally small, the differences between spatial and momentum components can be made as small as desired, up to a limit.

The application of the de Broglie equation can be used to set a lower limit on the size of a cell in state space:

$$\lambda = \frac{h}{p} = \frac{h}{(2mE)^{1/2}}$$

So

$$V_{i_{min}} = \lambda^3 = \frac{h^3}{(2mE)^{3/2}}$$

thus further increasing the probability of single state occupancy when the de Broglie condition is included in Equation 5-33. Under these conditions, the classical statistics may be applied where the de Broglie wavelength, based upon the average energy of the particles, is small compared to the average distance between particles. In the case of gases, conditions favoring this are those present at high temperatures and/or low densities.

The case requiring the fewest restrictions upon the particles in an assembly is illustrated by ideal gases. In this situation, the molecules are considered to be distinguishable and the Pauli condition of exclusion is not invoked; thus, more than one molecule may occupy the same state. The classical Maxwell-Boltzmann statistics are applicable in this case. Under the conditions of high temperatures and/or very low densities, as previously noted, there is very little probability of multiple occupancy of a state. So, at very low densities, real gas molecules would have a high probability of each occupying a single state (Equation 5-33). This is especially true when the lower limit is placed upon $V_{i_{min}}$.

However, multiple occupancy of a state is most probable for real gases under non-ideal conditions. The Maxwell-Boltzmann probability for the occupation of a state is given by

$$P = \frac{n_s}{g_s} = \frac{1}{\exp\left[(E_s - \bar{E})/k_B T\right]}$$

$$n_s \ll g_s$$

Here n_s is the number of particles, g_s is the number of states with energy E_s, and $\bar{E}$ is the average energy of the particles in the distribution. As E_s increases beyond $\bar{E}$ the probability of finding a particle in that state decreases. This is reasonable since fewer particles are expected the greater the difference in the energy of particles from the mean.

For the case in which $E_s \gg \bar{E}$, the Maxwell-Boltzmann function reduces to the approximate form

$$P = \frac{n_s}{g_s} \simeq \exp(-E_s/k_BT)$$

This also is known as the Boltzmann factor. It will be recalled that this expression was obtained earlier from the Einstein equation for average energy for cases where the exponential term is much greater than unity.

A more complex, nonclassical situation exists where the particles are indistinguishable, but may occupy the same state. This is the same as treating the particles as though the Heisenberg Uncertainty Principle applied, but the Pauli Exclusion Principle did not exist. Phonons and photons are among the particles (also called "bosons") which are treated in this way. The Bose-Einstein statistics are used for such particles. This approach gives the probability of occupancy of a state as

$$P = \frac{n_s}{g_s} = \frac{1}{\exp\left[(E_s - \mu)/k_BT\right] - 1}$$

Here μ is the chemical potential. In the case of the bosons of zero mass cited previously, $\mu = 0$ and the probability for this case is given by

$$P = \frac{n_s}{g_s} = \frac{1}{\exp(E_s/k_BT) - 1}$$

This expression, also known as the Planck distribution, was obtained earlier from the Einstein equation for average energy. It, in addition, may be reduced to the Boltzmann factor for certain conditions.

It will be noted that the probability of occupancy of a given state is higher when calculated using the Bose-Einstein relationship than when using the Maxwell-Boltzmann equation. This should be the case since the particles are indistinguishable in the Bose-Einstein case, a property that tends toward higher state occupancy in the absence of any restraints imposed by exclusion.

Some of the interrelationships between the expressions for the probable occupancy of a state as given by that derived from the Einstein expression for the average energy of an oscillator, the Maxwell-Boltzmann statistics and the Bose-Einstein statistics have been shown. Another interesting relationship exists between the Bose-Einstein and Fermi-Dirac statistics which are developed later in this chapter. It turns out that the chemical potential is very closely related to the Fermi energy. At 0 K, $\mu = E_F$ and for most metals the approximation $\mu \simeq E_F$ is very good because $E_F \gg k_BT$. Use is made of this relationship in dealing with phase equilibria in Chapter 7.

5.4.2. Quantum Mechanic Restrictions

It was shown previously (Chapter 3) that the quantum mechanics results in specific restrictions upon the behaviors of particles. Under these limitations, energy distributions are discrete. The Heinsenberg Uncertainty Principle is operative and does not permit an exact knowledge of the position and momentum of a particle. The particles must be indistinguishable because if they can be identified their properties are changed. In addition, the Pauli Exclusion Principle requires that no two particles can have the same set of quantum members (occupy the same "cell" in state space). The contrast between the two approaches is given in Table 5-3.

Table 5-3
COMPARISON OF PROPERTIES

Classical	Quantum
Particles identifiable	Particles indistinguishable
Continuous energy spectrum	Discrete energy spectrum
No limitations upon p and x or E and t	Uncertainty in p and x and E and t
No limitations upon the number of particles in a state	Definite limitations upon the number of particles in a state

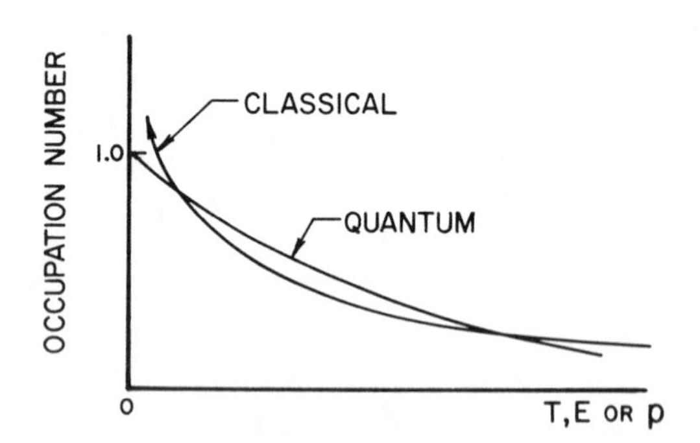

FIGURE 5-3. Differences in the probability of state occupation as given by classical and quantum statistics.

5.4.3. Indistinguishability

Both classical and quantum statistics provide means for the determination of a most probable distribution. Any distribution will be strongly influenced by the way in which its components are counted. For example, if the Pauli Exclusion Principle is operative, only one electron can occupy a given state. Such a state either will be occupied by one particle or it will be empty. This is one way of counting. Another basis for counting would be that in which several particles could occupy the same state.

The Maxwell-Boltzmann statistics are based upon being able to distinguish between particles. If several particles occupy the same state, they are not discriminated by the statistics and are counted. The quantum statistics differs in that although the particles are indistinguishable, only one particle can occupy a given state.

Using the Maxwell-Boltzmann statistics, since there are no restrictions upon the number of particles in a given state, large numbers of particles could occupy a few states. This could not occur when the quantum statistics are used. This difference in behavior is shown in Figure 5-3. The difference may be further understood by considering an assembly of particles in which two particles are to be in the lowest state. Using the Maxwell-Boltzmann approach, the combinations of two particles given by AB, AC, BC, etc. would all be different and would be counted as separate ways of filling the lowest state. In quantum statistics, however, these combinations of particles would be counted as only one way because the particles must be considered to be indistinguishable.

At higher temperature, energies, or momenta, multiple occupancy of a state is less likely to occur in the classical case. This results from the probability that the density of particles in a given state range is much smaller than at the low-energy ranges. Here, as previously noted, it is more probable that no two particles will occupy the same state. This accounts for the increasing similarities of the two statistics as the energy increases, Figure 5-3.

Table 5-4
SUMMARY OF STATISTICAL TREATMENTS

Type of particle	Example	Type of statistics
Distinguishable Unquantized	Ideal gas molecules	Maxwell-Boltzmann
Indistinguishable Unquantized	Real gas molecules at high temperature and/or low pressure	Maxwell-Boltzmann
Indistinguishable Quantized (zero or integral spin)	Phonons, photons, Cooper pairs, magnons, excitons	Bose-Einstein
Indistinguishable Quantized (half-integral spin)	Electrons, protons, neutrons	Fermi-Dirac

Modified from Levy, R. A., *Principles of Solid State Physics,* Academic Press, New York, 1968, 293. With permission.

The constraint of indistinguishability by itself cannot lead to a distribution different from that of Bose-Einstein statistics. This results from the fact that this can be adjusted to provide a high probability of one particle per state in a manner similar to that used to obtain Equation 5-32. The application of a lower limit to the cell size comparable to Equation 5-33 is also insufficient to provide a different distribution. Neither of these limitations necessarily ensures single occupancy. The additional constraint of exclusion is required to ensure that only single occupancy of a state occurs; this alone effectively guarantees that the Pauli Principle is operative.

5.4.4. Exclusion

The Pauli idea of exclusion can lead to two kinds of quantum statistics. This results from the way in which a state is defined. Those particles which require spin definition are more restricted than those which do not. They require that an additional "dimension" in state space be taken into consideration by the statistics.

It will be recalled that the concept of spin arose from the efforts to explain atomic spectra. It was assumed that an electron rotated on an axis about its center of mass and had both spin angular momentum and orbital angular momentum. The spin component is constant, $1/2 \cdot h/2\pi$ for a given direction. For the opposite direction, its spin is equal and opposite in sign. The necessity for this assumption was removed by Dirac (1928). This work showed that electrons must possess both spin and angular momenta when wave theory is reconciled with relativity theory.

Thus, particles possessing intrinsic spin must be treated statistically to include this property. Such particles must be treated differently from those which do not have intrinsic spin. The statistics for these must differ. Either a different statistical approach must be employed for such particles, or the effect of spin must be taken into account. Particles with nonintegral spin have an intrinsic angular momentum; examples of this include electrons, protons, and neutrons. Those with integral spin do not possess intrinsic angular momenta; phonons and photons illustrate this type of particle. Since such particles have no intrinsic angular momenta, spin is not considered in the Bose-Einstein statistics describing this behavior.

The statistics to be used for indistinguishable particles with intrinsic spin and which obey exclusion are subject to greater restrictions than are contained in either the Maxwell-Boltzmann or Bose-Einstein statistics. A state may be either singly occupied or empty. No other condition is permitted. The statistics must conform to this. The Fermi-Dirac statistics, derived in the next section, satisfy these constraints.

The conditions for the application of the various statistical approaches are summarized in Table 5-4.

The quantized, indistinguishable particles with integral spin are called "bosons". Systems composed of bosons have symmetric total wave functions. The quantized, indistinguishable particles with half-integral spin are known as "fermions". Systems composed of fermions have antisymmetric total wave functions (see Sections 3.11 in Chapter 3 and 5.7 in Chapter 5).

5.5. THE FERMI-DIRAC STATISTICS

A satisfactory quantum statistics must include the concepts of indistinguishability and exclusion. If spin is included by saying that two electrons of *opposite* spin can occupy the same state, then Equation 3-49 may be employed to describe an electron in terms of three quantum numbers. A "cell", or state, in state space is then defined by the quantum numbers n_x, n_y, and n_z. A particle in this state will have its energy given by

$$E_n = \frac{h^2}{8mL^2}[n_x^2 + n_y^2 + n_z^2] \quad (3\text{-}49)$$

Now consider a set of g_s degenerate states, each of which (by definition) has the same energy. In other words, in this case Equation 3-49 represents a small portion of the quasi-continuum of the energy. Let the number of electrons be n_s such that $n_s \ll g_s$. Of the g_s energy states, n_s are singly occupied and $(g_s - n_s)$ are empty. The first state may be filled in g_s ways; the second in $(g_s - 1)$ ways, the third in $(g_s - 2)$ ways, etc. There will be $(g_s - n_s + 1)$ ways of filling the states. However, some of these ways are indistinguishable, and distinguishable ways of counting the states are needed. Therefore, permutations must be excluded. This means that only the number of different ways that the indistinguishable electrons fill the distinguishable states must be determined. This can be accomplished by considering the number of combinations, W_s, of g_s taken n_s at a time, or

$$W_s = \frac{g_s!}{(g_s - n_s)!n_s!} \quad (5\text{-}34)$$

The probability of finding n_s particles in g_s states is

$$P(n_s) = P^{n_s}W_s = P^{n_s}\frac{g_s!}{(g_s - n_s)!n_s!} \quad (5\text{-}35)$$

This comes about from the fact that each of the states has the same probability of being filled. Since the probability of filling any one state is P, and there are n_s particles, the probability of filling n_s states is P^{n_s}.

For all energy levels, instead of just the small band of states originally considered, the probability of finding the particles is

$$P = P^N W_{s_1} W_{s_2} W_{s_3} \cdots = P^N W \quad (5\text{-}36)$$

where N is the total number of electrons and W is the product of all the W_{s_i}. The probability of finding a filled energy level, P^N, is a constant. The most probable arrangement is given by the maximum value for W. This can be obtained from the total number of combinations of electrons and states in the system in the following way:

$$W = \Pi_s \frac{g_s!}{n_s!\,(g_s - n_s)!} \tag{5-37}$$

It is more convenient to handle this as a sum rather than a product. A sum is obtained by taking the logarithm of both sides of Equation 5-37

$$\ell n\, W = \sum_s \left[\ell n\, g_s! - \ell n\, n_s! - \ell n (g_s - n_s)!\right] \tag{5-38}$$

Stirling's approximation may be used for this since large numbers are involved. This approximation is

$$\ell n\, a! \simeq a\, \ell n\, a - a$$

Thus, Equation 5-38 becomes

$$\ell n\, W = [\sum_s g_s \ell n g_s - g_s - n_s\, \ell n\, n_s + n_s - (g_s - n_s)\, \ell n (g_s - n_s) + (g_s - n_s)]$$

which reduces to

$$\ell n\, W = \sum_s [g_s \ell n g_s - n_s\, \ell n\, n_s - (g_s - n_s)\, \ell n (g_s - n_s)] \tag{5-39}$$

Taking the derivative, and recalling that g_s is a constant, gives

$$\partial \ell n W = \sum_s \left\{ \left[-n_s \cdot \frac{1}{n_s} - \ell n\, n_s\right] dn_s - \left[(g_s - n_s) \frac{-1}{(g_s - n_s)} - \ell n (g_s - n_s)\right] dn_s \right\}$$

which becomes

$$\partial \ell n W = \sum_s \left\{ \left[-1 - \ell n\, n_s\right] dn_s + \left[1 + \ell n (g_s - n_s)\right] dn_s \right\}$$

Collecting terms

$$\partial \ell n W = \sum_s \left[\ell n (g_s - n_s) - \ell n\, n_s\right] dn_s$$

$$= \sum_s \ell n \left[\frac{g_s - n_s}{n_s}\right] dn_s \tag{5-40}$$

It will be recalled that fixed conditions prevail. The total number of electrons is constant, or

$$\sum_s n_s = N \tag{5-41a}$$

and the total energy is constant:

$$\sum_s n_s E_s = E \tag{5-41b}$$

Undetermined constants can be chosen such that

$$\alpha \sum_s dn_s = 0 \tag{5-42a}$$

and

$$\beta \sum_s E_s\, dn_s = 0 \tag{5-42b}$$

Now equating Equation (5-40) to zero for the maximization of W, and including the undetermined constants 5-42a and 5-42b,

$$\partial \ell n W = \sum_s \left[\ell n \left(\frac{g_s - n_s}{n_s} \right) - \alpha - \beta E_s \right] dn_s = 0 \tag{5-43}$$

This equation can be satisfied only when

$$\ell n \frac{g_s - n_s}{n_s} - \alpha - \beta E_s = 0$$

This may be rewritten as

$$\ell n \frac{g_s - n_s}{n_s} = \alpha + \beta E_s$$

and expressed in exponential form by

$$\frac{g_s - n_s}{n_s} = \exp(\alpha + \beta E_s)$$

and rewritten as

$$\frac{g_s}{n_s} - 1 = \exp(\alpha + \beta E_s)$$

or

$$\frac{g_s}{n_s} = \exp(\alpha + \beta E_s) + 1 \tag{5-44}$$

Since the probability that a state will be occupied is given by n_s/g_s, the basic Fermi-Dirac equation is given by the reciprocal of Equation 5-44, or

$$\frac{n_s}{g_s} = \frac{1}{e^{\alpha + \beta E_s} + 1} \tag{5-45}$$

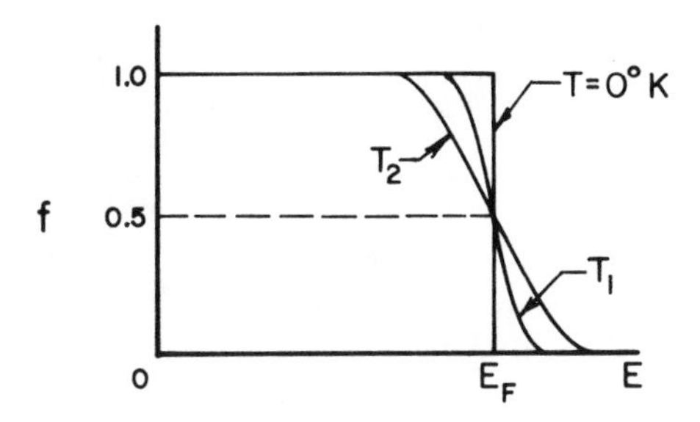

FIGURE 5-4. The effect of temperature upon the Fermi-Dirac function; $T_2 > T_1 \gg 0K$.

At very high energies $n_s/g_s \ll 1$, so, as described previously in Section 5.4.1,

$$n_s \simeq g_s\, e^{-\beta E_s} \tag{5-46}$$

This is the same as the previously noted Boltzmann factor. By comparison,

$$\beta = \frac{1}{k_B T} \tag{5-47}$$

Then the Fermi-Dirac function becomes

$$f = \frac{n_s}{g_s} = \frac{1}{e^{\alpha + E/k_B T} + 1} \tag{5-48}$$

An energy, known as the Fermi energy, can be defined such that

$$\alpha = -\frac{E_F}{k_B T} \tag{5-49}$$

This relationship also may be derived by the use of statistical mechanics. It will be recalled that the Fermi energy is closely related to the chemical potential. Then the Fermi-Dirac function becomes, upon substitution of Equation 5-49 into 5-48,

$$f = \frac{1}{e^{(E - E_F)/k_B T} + 1} = f(E,T) \tag{5-50}$$

This gives the probability of single occupancy of a quantum state by an electron.

This equation must satisfy the conditions given by Equation 5-29 for $T = 0$. For $E < E_F$, the denominator of Equation 5-50 equals $e^{-\infty} + 1 = 1$. Then $f = 1$. For $E > E_F$, the denominator becomes infinite and $f = 0$. Thus, these important, prior conditions have been satisfied.

It also will be noted that for any temperature above 0 K where $E = E_F$, the denominator of Equation 5-50 equals $e^{o} + 1 = 2$, and $f = 1/2$. Since E_F is virtually unaffected by temperature (Equation 5-26), the Fermi level is that energy at which the probability of state occupancy by an electron is exactly 0.5. The effect of temperature upon the Fermi-Dirac function is shown in Figure 5-4.

It was shown in a qualitative way in Section 5.2 that only electrons near the top of the well can enter into physical processes. It is now possible to state this more exactly: only electrons within a small energy range about E_F can participate in these processes. Consider the behavior of Equation 5-50 as a function of temperature. Where $(E - E_F)$ equals $-2\ k_BT$, $f = 0.88$. This means that the probability that a state will be occupied is 0.88. States with energies farther below E_F will possess increasingly higher probabilities of occupancy. There is little chance that electrons from states lower than these can be promoted into such levels because of the high probability of their prior occupancy. States in the neighborhood of E_F have about 0.5 probability of accepting an electron from a lower level. Where $(E - E_F)$ equals $+2\ k_BT$, the probability of the occupancy of a state by an electron is only 0.12. These states have a high probability (0.88) of availability for occupancy by an electron. Electrons promoted from lower levels by normal energies are accepted readily in such states. States beyond $E_F + 2\ k_BT$ also are available, but require higher than normal energies for occupancy. From this it can be seen that only a small fraction of the valence electrons are available for engaging in a physical process. This constitutes the answer to the second question posed at the beginning of Section 5.2: only a small percentage of the electrons, those with energies near E_F, can be involved. This is determined by the Fermi-Dirac function.

Where only those electrons with energies very much greater than E_F are being considered, the exponential term in Equation 5-50 becomes much greater than unity and the Fermi-Dirac function is approximated by

$$f \simeq \frac{1}{\exp\left[(E - E_F)/k_BT\right]} \qquad (5\text{-}50a)$$

This is of the same form as the Maxwell-Boltzmann equation and in the high-energy range, accordingly, is named the Boltzmann tail. The Bose-Einstein function gives identical results when $E_s \gg \mu$ and, as discussed previously, $\mu \simeq E_F$. Such high-energy electrons normally are not of sufficient number to require consideration in explanations of the properties of normal metals. In the case of intrinsic semiconductors, the small number of electrons in the conduction band must be taken into account by Equation 5-50a to explain their properties (see Section 11.1 in Chapter 11).

5.6. FERMI-SOMMERFELD THEORY OF METALS

The Sommerfeld theory provides a means for computing the density of states (Equation 5-21). The Fermi-Dirac function (Equation 5-50) shows how the electrons are distributed in those states as well as which of them can enter into physical processes. These equations along with Equation 5-28 can now be used to show their combined effects graphically, as in Figure 5-5.

It is apparent from the preceding discussion and Figure 5-5c that only relatively few electrons in states near E_F are excited, or promoted, by the thermal energy to states above E_F. This will be illustrated further in the next section.

Other ways of showing this are given in Figure 5-6a in terms of momentum and in Figure 5-6b in terms of a well.

In Figure 5-6a, given for momentum space, the ripples are the momenta equivalents of less than $\pm 2\ k_BT$, or $\pm(2\ mk_BT)^{1/2}$. The well picture (Figure 5-6b) shows little "waves" on the Fermi "sea". In both cases the variations are within the order of $\pm 2\ k_BT$. The ratios of the wavy areas to the total areas are intended to give a visual approximation of the small numbers of electrons which are made available by thermal activation.

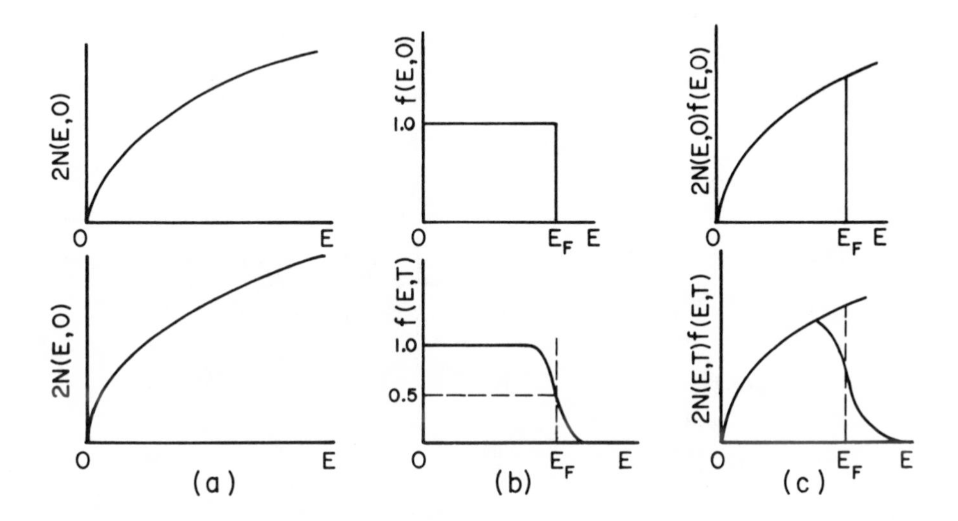

FIGURE 5-5. Fermi-Sommerfeld distribution of electrons. (a) Sommerfeld distributions; (b) Fermi-Dirac functions; (c) products of (a) and (b) at 0 and T K.

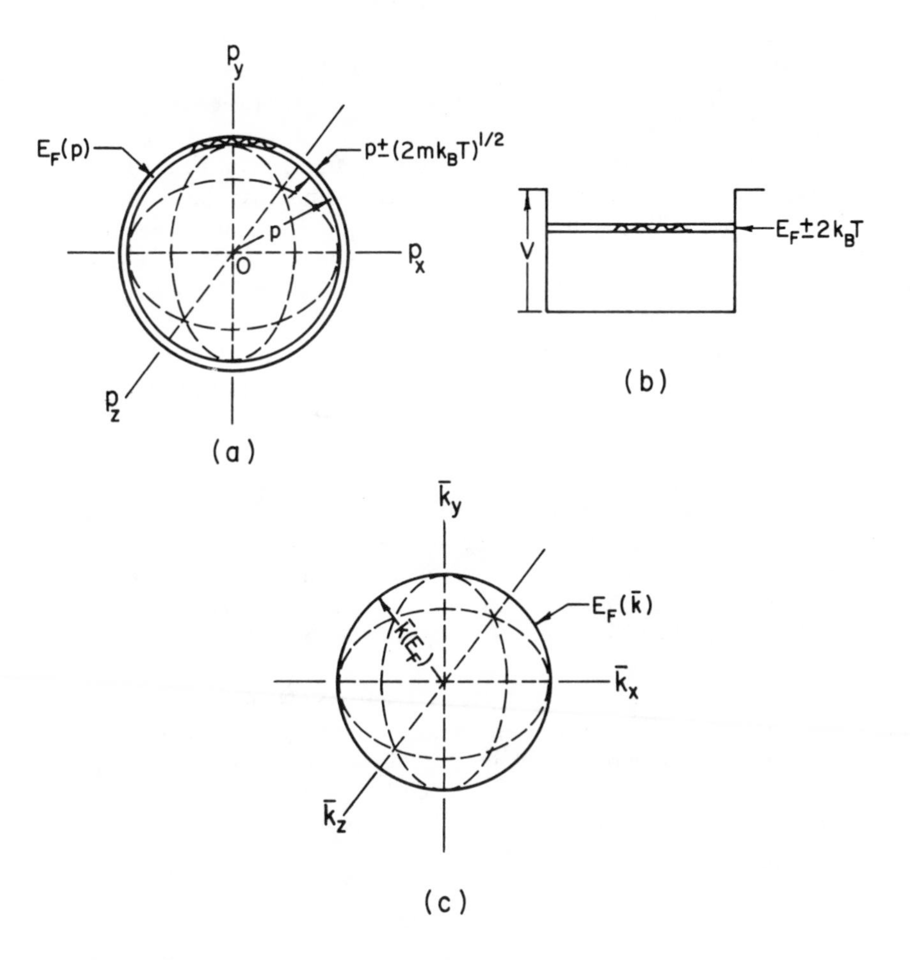

FIGURE 5-6. The effect of temperature on the Fermi surface. (a) in momentum space; (b) in a potential well (after C. W. Curtis, with permission); (c) the Fermi surface in $\bar{k}$ space.

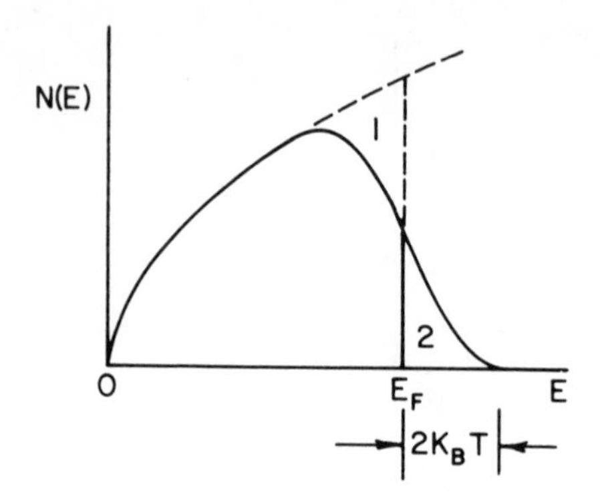

FIGURE 5-7. Curve of the density of states used for the approximation of the electron heat capacity of metals.

Another useful way of representing electron states is shown in Figure 5-6c for nearly free electrons. Here the states occupied up to the Fermi level are shown as a spherical surface in wave-vector space; this is known as the Fermi surface. This is a three-dimensional plot of Equation 3-26. Since energy and momentum are related, $E = p^2/2m$, Figures 5-6a and 5-6c are different ways of expressing the same behavior. Use is made of this correspondence in Figure 5-10a, a two-dimensional plot in momentum space. The vector p shown in this figure is the momentum of all electrons occupying states at the Fermi surface.

More complex representations, based upon this approach, can be used to illustrate the basis for the anisotropic properties of crystals. In this case the influences of the bounding surfaces of the Brillouin zone have important effects upon the properties of the electrons. They may no longer be considered as being nearly free; Equation 3-26 does not hold under these conditions and the shape of the Fermi surface becomes more complicated.

Despite these complications, a model which makes use of a spherical Fermi surface can be used to provide an elementary, but somewhat oversimplified approach to the theory of alloy phases (see Section 10.2.3.1 and the electron phases discussed in Section 10.6.6, Chapter 10, Volume 3).

5.6.1. Heat Capacity of Normal Metals

The combined results of the Fermi-Dirac and Sommerfeld theories will be used to examine the heat capacity of normal metals, metals with completed inner levels and partially filled, outer, valence states. If temperatures greater than θ_D are considered, the model may be simplified for purposes of approximation as shown, greatly exaggerated, in Figure 5-7.

That fraction, f(N), of the total number of electrons, N, which can be thermally excited are promoted from area 1 to states in area 2 in the figure. These electrons have energies up to about $2k_BT$ above E_F. Since $2k_BT$ is very small with respect to E_F, the slope of the "hypotenuse" of area 2 actually is very steep. Thus, the approximations may be made, in terms of the energy ranges involved, that

$$f(N) \simeq \frac{3}{4} \cdot \frac{2k_BT}{E_F} = \frac{3}{2} \cdot \frac{k_BT}{E_F}$$

and that the number of thermally excited electrons is

$$N(T) = Nf(N) \simeq \frac{3Nk_BT}{2E_F}$$

Those electrons which have been excited to energies equal to or greater than E_F may be treated as being nearly free (Section 5.8). The average energy of each such electron is close to $3/2\ k_BT$. Therefore, the total in the thermal energy of the electrons is given by the product of the number of excited electrons and the average energy of one of these electrons, or

$$U_{e,T} \simeq N(T) \cdot \frac{3}{2} k_BT \simeq \frac{3Nk_BT}{2E_F} \cdot \frac{3}{2} k_BT = \frac{9}{4} \frac{Nk_B{}^2T^2}{E_F}$$

Here $U_{e,T}$ approximates the electron contribution to the internal energy of a solid which is caused by thermal excitation. From this the heat capacity of the electrons is found to be

$$C_{V,e} = \frac{\partial U_{e,T}}{\partial T} \simeq \frac{9}{2} \frac{Nk_B{}^2T}{E_F} \qquad (5\text{-}51)$$

The total internal energy of the electrons is

$$U_e = U_{e,0} + U_{e,T}$$

in which $U_{e,0}$ is the energy of the electrons at 0 K. Similarly, the total internal energy of the ion cores is

$$U_i = U_{i,0} + U_{i,T}$$

And the total internal energy of the metal is

$$U = U_i + U_e = U_{i,0} + U_{i,T} + U_{e,0} + U_{e,T}$$

The zero-point energies of the ion cores and of the electrons are constants. These quantities vanish upon differentiation with respect to temperature and the heat capacity of the metal is given by

$$C_V = \frac{\partial U}{\partial T} = \frac{\partial U_{i,T}}{\partial T} + \frac{\partial U_{e,T}}{\partial T} = 3Nk_B + \frac{9}{2} \frac{Nk_B{}^2T}{E_F} \qquad (5\text{-}52)$$

Equation 5-52 is the sum of the heat capacities of the ions and the electrons and may be written as

$$C_V = C_{V,i} + C_{V,e} = 3Nk_B + \frac{9}{2} Nk_B \cdot \frac{k_BT}{E_F} \qquad (5\text{-}52a)$$

Then, by dividing through by the ionic portion,

$$\frac{C_V}{C_{V,i}} = 1 + \frac{C_{V,e}}{C_{V,i}} = 1 + \frac{3}{2} \frac{k_BT}{E_F} \qquad (5\text{-}52b)$$

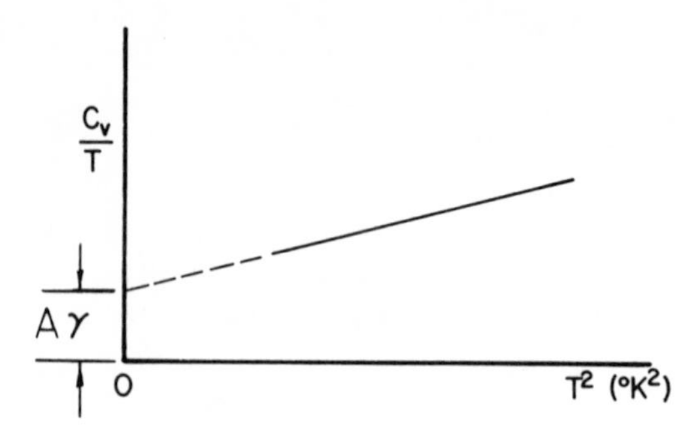

FIGURE 5-8. Electron heat capacity derived from heat capacity data for metals at very low temperatures.

Equation 5-52b may be used to approximate the relative contribution of the electrons to the heat capacity of a metal. For example, if k_BT is taken to be about 0.03 eV near room temperature, and E_F as being about 6 eV, then the electron contribution to the heat capacity of a metal is about 1.5% of the total. This gives $C_V \simeq 3R$ for $T \simeq \theta_D$ and imparts a small, additional, positive slope to C_V as a function of temperature, in agreement with experiment.

A more accurate calculation by Seitz gives the electron contribution to the heat capacity of a metal as

$$C_{V,e} = \frac{\pi^2}{2}\frac{Nk_B^2T}{E_F} \tag{5-53a}$$

It will be recognized that the coefficient 9/2 in Equation 5-51 is very close to $\pi^2/2$ in Equation 5-53a. In both cases the small electron contribution to the heat capacity eliminates the inherent inconsistency of the Drude-Lorentz model (Section 5.1.3). The Drude-Lorentz model included all of the valence electrons. The model employed here considered only those valence electrons in the energy range within $2k_BT$ above E_F; these, as represented by f(N) above, constitute only about 1.5% of the total number of electrons.

Both Equations 5-51 and 5-53a are based upon relatively simple models and both involve approximations. In order to account for a more general model, since the coefficient is different for each metal, the electron contribution is best given by

$$C_{V,e} = A\,\frac{\pi^2}{2}\frac{Nk_B^2T}{E_F} = A\gamma T \tag{5-53b}$$

in which A is a constant for a given elemental metal which is not a transition element. However, small variations in A must be taken into consideration to account for variations in the electron heat capacities and in the thermoelectric properties of alloys of the given metal. Thus, A has been termed the conduction coefficient (see Sections 5.6.2, 7.8.4 and 7.8.6 in Chapter 7, Volume II, and subsequent sections).

Equation 5-53b is also useful at very low temperatures. It will be recalled from the Debye "T^3 law" (Equation 4-111 in Chapter 4) that $C_{V,i}$ varies as T^3 at low temperatures. $C_{V,e}$ has a linear dependence upon T. Thus, for very low temperature, the heat capacity of a metal may be expressed as

$$C_V = A\gamma T + BT^3 \tag{5-54}$$

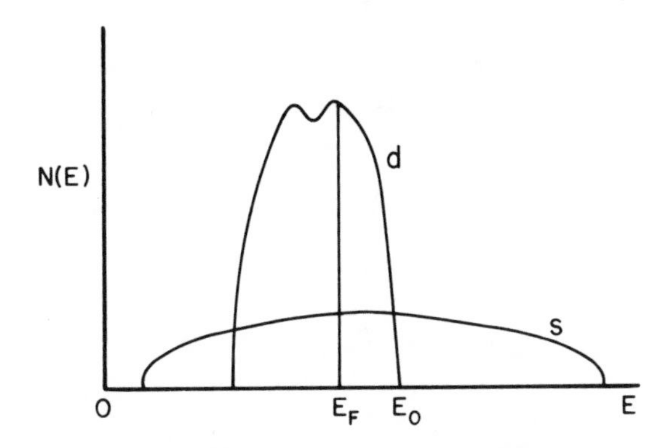

FIGURE 5-9. Schematic diagram of band overlap in a transition element.

Heat capacity measurements, made at very low temperatures, have been used to provide greater understanding than Equation 5-53b as it is written above. A plot of C_V/T versus T^2 gives $A\gamma$ as the intercept as is shown schematically in Figure 5-8.

The intercepts can be significantly different from $\pi^2 N k_B^2/2E_F$ as given by Equation 5-53a; in reality the intercepts are $A\gamma = A\pi^2 N k_B^2/2E_F$. This is a result of the simplifying assumption of nearly free electrons which was used to obtain Equation 5-51. The conduction coefficients A, thus, are correction factors for γ; this explains the necessity for their inclusion in Equation 5-53b. This is discussed more fully in terms of the thermal effective mass of the electrons in a metal in Section 5.6.2.

While the electronic contribution to the heat capacity is shown to be small, it comprises an appreciable fraction of the heat capacity at very low temperatures; it, therefore, must be considered in this region. This can be of importance in cryogenic applications.

5.6.2. Heat Capacity of Transition Elements

The preceding discussions were concerned with normal metals. The transition elements are characterized by incomplete, inner d levels as well as partially filled, outer s levels in the solid state. Some of the transition elements behave as though the absence of electrons in the d band was the same as the presence of positive charges entering into the conduction process. Such positive charges are known as "holes".

Both the d and s levels overlap. This must be taken into account to explain the properties of such elements. A highly oversimplified sketch of the density of states of a transition element showing this overlap is given in Figure 5-9. This is discussed further in the section on band theory (5.7).

In the case of the normal metals the states were described by the way in which they were filled, starting with the lowest energy level. An equally suitable model can be based upon the way in which such states remain unfilled. This is the approach used for transition elements. When this is followed, the electron contribution to the heat capacity is found to be

$$C_{V,e} = A' \frac{\pi^2}{6} \frac{N k_B^2 T}{(E_o - E_F)} = A'\gamma' T \tag{5-55}$$

in which the coefficient A' accounts for the specific electron behavior in a given element or alloy and E_o is the energy of the top of the d band.

The overlap results in mixed s-d behavior and is sometimes considered to be hybrid-

Table 5-5
EXPERIMENTAL AND FREE ELECTRON VALUES OF ELECTRONIC HEAT CONSTANT γ OF METALS

Li	Be											B	C	N
1.63	0.17													
0.749	0.500													
2.18	0.34													
Na	Mg											Al	Si	P
1.38	1.3	Observed γ in mJ mol^{-1} K^{-2}										1.35		
1.094	0.992	Calculated free electron γ in mJ mol^{-1} K^{-2}										0.912		
1.26	1.3	m_{th}/m = (observed γ)/(free electron γ)										1.48		
K	Ca	Sc	Ti	V	Cr	Mn (γ)	Fe	Co	Ni	Cu	Zn	Ga	Ge	As
2.08	2.9	10.7	3.35	9.26	1.40	9.20	4.98	4.73	7.02	0.695	0.64	0.596		0.19
1.668	1.511									0.505	0.753	1.025		
1.25	1.9									1.38	0.85	0.58		
Rb	Sr	Y	Zr	Nb	Mo	Tc	Ru	Rh	Pd	Ag	Cd	In	Sn (w)	Sb
2.41	3.6	10.2	2.80	7.79	2.0	---	3.3	4.9	9.42	0.646	0.688	1.69	1.78	0.11
1.911	1.790									0.645	0.948	1.233	1.410	
1.26	2.0									1.00	0.73	1.37	1.26	
Cs	Ba	La	Hf	Ta	W	Re	Os	Ir	Pt	Au	Hg (α)	Tl	Pb	Bi
3.20	2.7	10.0	2.16	5.9	1.3	2.3	2.4	3.1	6.8	0.729	1.79	1.47	2.98	0.008
2.238	1.937									0.642	0.952	1.29	1.509	
1.43	1.4									1.14	1.88	1.14	1.97	

From compilations kindly furnished by N. Phillips and N. Pearlman to C. Kittel. From Kittel, C., *Introduction to Solid State Physics*, 5th ed., John Wiley & Sons, New York, 1976, 212. With permission.

ized, or mixed, behavior. This is used to account for such other electron transport properties as thermal, magnetic, and thermoelectric behaviors.

It was noted in Section 5.6.1 that the conduction coefficient A of Equation 5-53b is required to account for variations in the conduction mechanisms of each element. The same is true for A′ of Equation 5-55. The necessity for these factors arises because these equations are based upon nearly free electrons. This implies that the electron mass is assumed to be that of a free electron. Corrections must be made for this simplification since the electron mass is influenced by electron-electron effects (Section 5.6 and 5.7), electron-phonon effects (Section 5.6.3, and Sections 6.3 and 6.4 in Chapter 6) and by the periodic potential of the lattice (Sections 5.7 and 5.8.4, and Section 11.1.2 in Chapter 11). The coefficients A and A′ represent the combined effects of these reactions upon γ and γ'. Both coefficients are functions of the ratio of the thermal effective mass, m_{th}, to the mass, m, of the free electrons. The conduction coefficients are given by

$$A \text{ or } A' = \frac{m_{th}}{m} = \frac{\gamma\,(\text{observed})}{\gamma\,(\text{free})}$$

Data for A are given in terms of m_{th}/m in Table 5-5 along with experimental and calculated values for γ. A′ may be approximated from the data in this table by using

$$\gamma'\,(\text{free}) \simeq \pi^2 N k_B^{\,2}/6(E_0 - E_F)$$

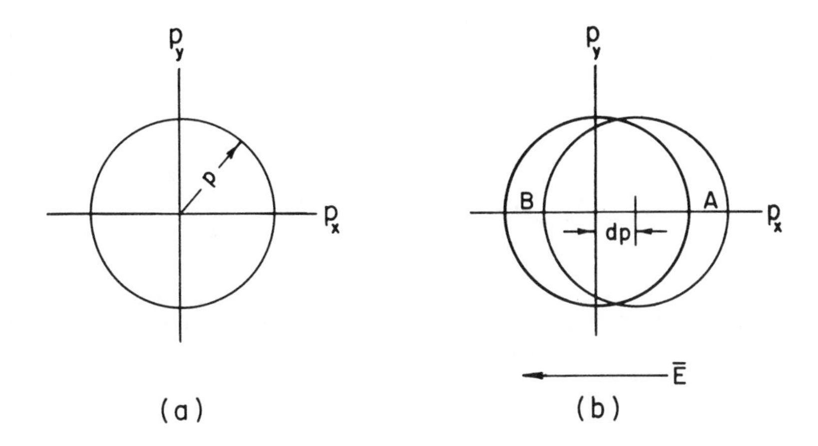

FIGURE 5-10. Effect of an applied electric field upon the momentum states of a metal.

in conjunction with the γ' (observed) data which are given for the transition elements.

The presence of alloying elements in solid solution in a given metal will affect the value of A or A′. Small changes of this kind have important influences upon thermoelectric behavior (see Section 7.8.5 in Chapter 7).

5.6.3. Electrical Conductivity

The electrical conductivity of a normal metal will be discussed in terms of the way in which the momenta of electrons at or near E_F are affected by the applied field. It is approached by considering the effects of an applied electric field, $\bar{E}$, upon the momentum states in a solid, as shown in Figure 5-10. This is based upon the model shown in Figure 5-6a.

The field causes a slight net shift in the momenta of electrons to slightly higher levels. Those electrons at the high levels, lune A, are scattered, give off a phonon, and drop back to the lower levels at or near lune B. This process results in a small net flow of electrons. The asymmetry of the momentum space can be approximated by the shift of the center of the momentum sphere to some position dp. The drift velocity induced by the electric field is small compared to the average velocity, so dp is small. The electrons are considered to behave elastically, since no scattering mechanism is considered.

The force acting on an electron is, from Equations 3-4 and 5-2

$$F = ma = \frac{\partial p}{\partial t} = e\bar{E} \tag{3-42}$$

By means of the de Broglie relationship and the expression for wave vector, one obtains

$$p = \frac{h}{\lambda} = \frac{h}{2\pi} \cdot \frac{2\pi}{\lambda} = \frac{h}{2\pi} \bar{k} \tag{5-56}$$

which, upon differentiation, gives

$$\frac{dp}{dt} = \frac{h}{2\pi} \frac{\partial \bar{k}}{dt} \tag{5-57}$$

The substitution of Equation 5-57 into Equation 3-42 results in

$$F = \frac{\partial p}{\partial t} = \frac{h}{2\pi}\frac{\partial \bar{k}}{\partial t} = \bar{E}e$$

from which

$$\partial \bar{k} = \frac{2\pi e \bar{E}}{h} \partial t \tag{5-58}$$

This represents the displacement of the center of the Fermi sphere in $\bar{k}$ space as a result of the application of the electric field. The shift of the center of the sphere in p space (Figure 5-10) also may be determined from the differentiation of Equation 5-56

$$\partial p = \frac{h}{2\pi} \partial \bar{k} \tag{5-59}$$

and the substitution of Equation 5-58 to give

$$\partial p = \frac{h}{2\pi}\frac{2\pi e \bar{E}}{h} \partial t = e\bar{E}\partial t \tag{5-60}$$

Actually, the electron interacts with phonons, impurities, imperfections, etc. in the lattice and is scattered. If the time between such interactions is $\tau(E_F)$, the relaxation time of electrons with $E \geqslant E_F$, and the change in the velocity (drift velocity) of the electron is ΔV, then from Equations 3-3a and 5-60,

$$\Delta V = \frac{\Delta p}{m} \approx \frac{e\bar{E}\tau(E_F)}{m} \tag{5-61}$$

where m is the effective mass of the electron and $\tau(E_F) \sim \partial t$. Since $\tau(E_F)$ is of the order of 10^{-14} seconds, this approximation is fair. Equation 5-61 leads directly to the mobility, μ, which is defined as the drift velocity per unit field, or $\mu \equiv \Delta V/\bar{E} = e\tau(E_F)/m$.

The current density is given, in terms of the number of electrons per unit volume with $E \geqslant E_F$, $n(E_F)$, as

$$j = n(E_F)e\Delta V$$

then using Equation 5-61

$$j = \frac{n(E_F)e^2\bar{E}\tau(E_F)}{m} \tag{5-62}$$

The electrical conductivity is determined from Ohm's Law as

$$j = \sigma\bar{E} = \frac{n(E_F)e^2\bar{E}\tau(E_F)}{m}$$

or, since $\bar{E}$ vanishes,

$$\sigma = \frac{n(E_F)e^2\tau(E_F)}{m} = \frac{1}{\rho} \tag{5-63}$$

Table 5-6
ELECTRICAL RESISTIVITY AND THERMAL CONDUCTIVITY OF SOME ELEMENTS NEAR ROOM TEMPERATURE (20°C)[14]

Element	κ^a	ϱ^b
Aluminum	0.53	2.6548
Antimony	0.045	39.0
Beryllium	0.35	4
Bismuth	0.020	106.8
Cadmium	0.22	6.83(0C)
Calcium	0.3	3.91(0C)
Carbon (Graphite)	0.057	13.75(0C)
Chromium	0.16	12.9(0C)
Cobalt	0.165	6.24
Copper	0.941	1.6730
Germanium (impure)	0.14	46
Gold	0.71	2.35
Indium	0.057	8.37
Iodine	1.04×10^{-4}	1.3×10^{15}
Iridium	0.14	5.3
Iron	0.18(0C)	9.71
Lead	0.083(0C)	20.648
Lithium	17	8.55(0C)
Magnesium	0.367	4.45
Mercury	0.0196(0C)	98.4(50C)
Molybdenum	0.34	5.2(0C)
Nickel	0.22(25C)	6.84
Palladium	0.168(18C)	10.8
Platinum	0.165(17C)	10.6
Plutonium	0.020(25C)	141.4(107C)
Rhenium	0.17	19.3
Rhodium	0.21(17C)	4.51
Silicon (impure)	0.20	10(0C)
Silver	1.0(0C)	1.59
Sodium	0.32	4.2(0C)
Sulfur (yellow)	6.31×10^{-4}	2×10^{23}
Tantalum	0.130	12.45(25C)
Tellurium	0.014	4.35×10^5(23C)
Thallium	0.093	18(0C)
Thorium	0.090(100C)	13(0C)
Tin	0.150(0C)	11(0C)
Titanium		42
Tungsten	0.397(0C)	5.65(27C)
Uranium	0.150(0C)	11(0C)
Zinc	0.27(25C)	5.916

[a] Cal/cm sec deg.
[b] Ohm-cm $\times 10^6$.

Some typical values of the electrical resistivity of elements are given in Table 5-6.

Where $\tau(E_F)$ is taken as

$$\tau(E_F) = \frac{L(E_F)}{V(E_F)} \qquad (5\text{-}64)$$

in which $L(E_F)$ is the distance travelled by an electron between collisions and $V(E_F)$ is its average velocity, this may be used in Equation 5-63 to obtain another expression for the electrical conductivity as

$$\sigma(E_F) = \frac{n(E_F)e^2 L(E_F)}{m\,V(E_F)} \tag{5-63a}$$

This essentially is the same as the reciprocal of Equation 5-6, since $V(E_F)$ was defined as the average velocity of an electron engaged in the conduction process. The magnitude of $V(E_F)$ is about 10^8 cm/sec. Here the mean free path is found to be of the order of about 10^2 interionic spacings, whereas in the Drude-Lorentz theory it is of the order of the interionic distance. Another significant difference between these two equations lies in the way in which the numbers of electrons involved are computed in each case.

5.6.4. Thermal Conductivity

A quantum-mechanic expression for the electron contribution to the thermal conductivity of a metal can be obtained in the same way as the expression (Equation 4-151) derived for insulators. However, the factors involved in the present case refer to the electrons and not to the phonons. Thus,

$$\kappa = \frac{1}{3} C_{V,e}\, V(E_F) L(E_F) \tag{5-65}$$

where

$$C_{V,e} = \frac{\pi^2}{2} \cdot \frac{N k_B{}^2 T}{E_F} \tag{5-53a}$$

Thus, κ is expressed as

$$\kappa = \frac{1}{3} \cdot \frac{\pi^2}{2} \cdot \frac{N k_B{}^2 T}{E_F} V(E_F) L(E_F) \tag{5-66}$$

Some typical values for κ are given in Table 5-6.

5.6.5. Wiedemann-Franz Ratio

The Wiedemann-Franz ratio now may be determined from Equations 5-66 and 5-63a. Starting with the classical expression for energy, where m is the effective mass,

$$E_F = \frac{1}{2} m [V(E_F)]^2 \tag{5-67}$$

Equation 5-66 becomes

$$\kappa = \frac{1}{3} \cdot \frac{\pi^2}{2} \cdot N k_B{}^2 T \cdot \frac{2}{m[V(E_F)]^2} \cdot V(E_F) L(E_F)$$

or, simplifying,

$$\kappa = \frac{\pi^2}{3m} \cdot N k_B{}^2 T \cdot \frac{L(E_F)}{V(E_F)} \tag{5-68}$$

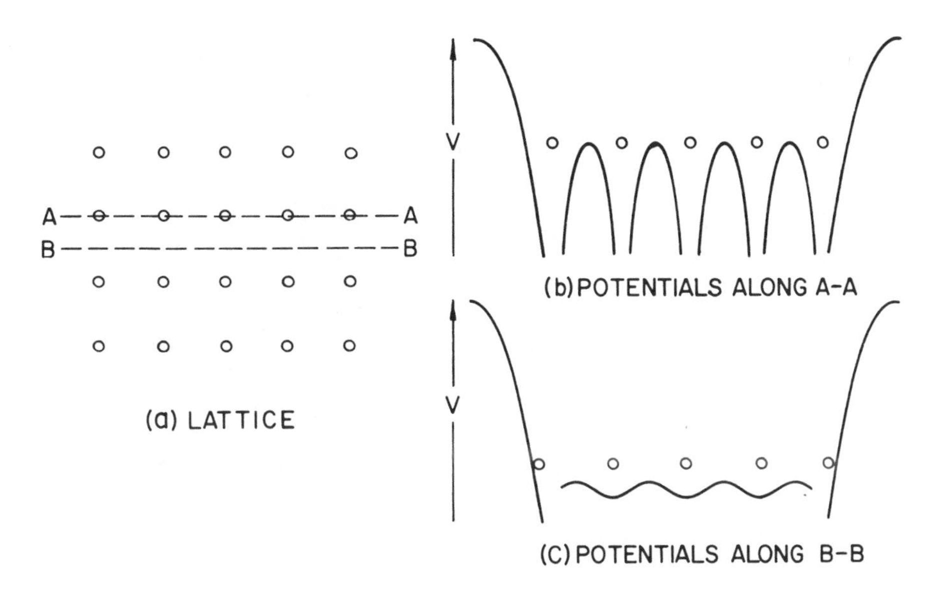

FIGURE 5-11. Schematic diagram of periodic potentials in a lattice.

Upon division of Equation 5-68 by Equation 5-63a, the Wiedemann-Franz ratio is obtained:

$$\frac{\kappa}{\sigma} = \frac{\pi^2}{3m} N k_B^2 T \cdot \frac{L(E_F)}{V(E_F)} \cdot \frac{m\,V(E_F)}{n\,e^2 L(E_F)}$$

and

$$\frac{\kappa}{\sigma T} = \frac{\pi^2}{3}\left[\frac{k_B}{e}\right]^2 = L_o \qquad (5\text{-}69)$$

The Lorentz number given by Equation 5-69 is closer to experiment than that provided by Equation 5-14. The slightly better fit comes from the larger coefficient of the former. Some values for experimentally determined Lorentz numbers are given in Table 5-1. Much greater variations than these occur at very low temperatures where L_o approaches zero approximately as T^2. In the case of alloys, L_o deviates widely, but systematically, from this as a function of composition and temperature for each alloy system.

5.6.6. Summary of Fermi-Sommerfeld Theory

The Fermi-Sommerfeld theory removes the dilemma encountered in the Drude-Lorentz approach. The electron contribution to the heat capacity is approximately correct and has the proper temperature dependence. The expression for the electrical conductivity is correct, but the relaxation time is undefined. The Wiedemann-Franz ratio is slightly better than that given by the Drude-Lorentz theory.

The major difficulty with this theory is that it provides no scattering mechanism for the electrons. This means that there is no direct way to calculate either $\tau(E_F)$ or $L(E_F)$. These factors cannot be calculated from first principles.

In addition, this theory is intended to explain the properties of metals; it does not purport to provide a basis for the classification of other types of solids on the basis of their electronic structure. However, when the Fermi-Sommerfeld concepts are used in

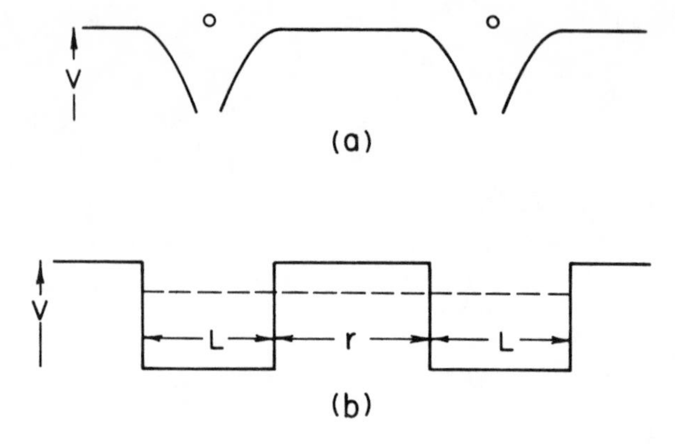

FIGURE 5-12. (a) Schematic diagram of potentials between two ions; (b) square-well simulation.

conjunction with band theory, the combination provides another means for the visualization of the essential differences between the principle classes of solids (see Section 5.7.1).

5.7. BAND THEORY OF SOLIDS

The foregoing theories have considered the electrons within the metal to be in a constant internal potential as shown in Figure 3-2, Chapter 3. However, the potential within a metallic lattice is not constant. Each ion within the lattice constitutes a potential well (Section 3.4, Chapter 3). This is shown schematically in Figure 5-11a. This periodicity of the potentials gives rise to the band model of solids.

It will be noted that this model neglects electron-electron interactions. This problem has been eliminated by the concept, introduced by Landau, that quasi-particles are involved rather than electrons. These quasi-particles consist of a central electron surrounded by a volume lower in electron density, known as a correlation hole. This, in turn, is encompassed by a volume higher in electron density and consists of electrons electrostatically repelled from the central electron which move around it in a way similar to that of a liquid flowing around a moving object. Electrons rather than quasi-particles will be considered here for simplicity.

Two wells of the type shown in Figure 3-8a, Chapter 3, very widely separated, will be brought closely together. This is the same as bringing together two widely separated ions. This situation can be approximated by bringing two widely separated hydrogen ions together as in Figure 5-12. Here the ions are considered to be motionless, since their motions are small compared to those of the electrons. If an electron is introduced into this two-well system, the resulting quantum-mechanic effects are indicated in Figure 5-13a. The energy in each well is given by Equation 3-40 for n = 1, or $E_1 = h^2/8mL^2$. The wave functions are designated by two subscripts; the first describes the state and the second, the well. Here ψ_{11} and ψ_{12} are identical within the wells. They quickly taper off and become very small between the wells. The wave function for the system is

$$\psi = \psi_{11} + \psi_{12} \tag{5-70}$$

Since ψ_{11} and ψ_{12} are identical,

$$\psi_{11}{}^{*}\psi_{11} \equiv \psi_{12}{}^{*}\psi_{12} \tag{5-71}$$

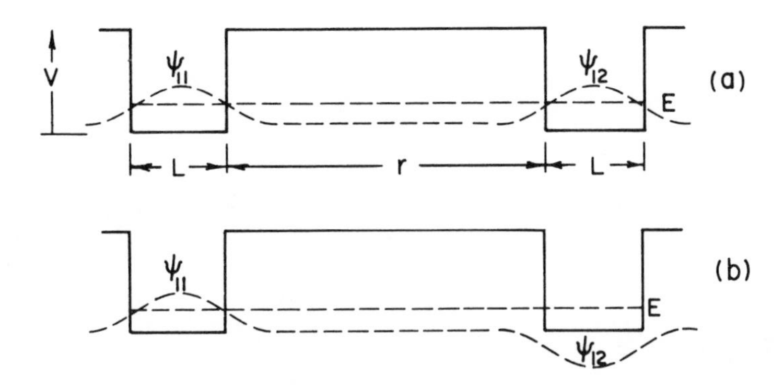

FIGURE 5-13. Schematic diagrams of (a) symmetric and (b) antisymmetric wave functions in a two-well system. (Modified from Sproull, R. L., *Modern Physics*, John Wiley & Sons, New York, 1956, 205. With permission.)

It is, therefore, equally likely that the electron will be found in either well. The probability densities become very small between the wells. Therefore, there is a negligible probability of finding the electron between the wells and an equal probability of finding it in either well.

As shown in Figure 5-13b, another solution of Schrödinger's equation is also possible. ψ_{12} reverses its sign when the electron changes its spatial coordinates in going from one well to the other and is said to be antisymmetric (see Section 3.11, Chapter 3). However, their shapes are identical, and Equations 5-70 and 5-71 are valid. The probability density of finding the electron remains the same as in the symmetric case. In addition, the energies also remain unchanged. Equation 3-40 gives

$$E_1 = \frac{h^2}{8mL^2} = E_S = E_A \; ; \; n = 1 \qquad (5\text{-}72)$$

where E_S is the energy in the symmetric case and E_A that of the antisymmetric case.

Interesting and significant changes take place in the wave functions as the wells are brought closer together until they finally abut. An attempt to describe these changes with decreasing separation, r, of the wells is shown in Figure 5-14.

As the wells are brought closer together, the probability of finding the electron between the wells, in the symmetric case, becomes greater. At the position where the wells abut, the greatest probability of finding the electron is at the common wall when the wells abut. In the antisymmetric case the probability of finding the electron is zero at the common wall when the wells abut.

The symmetric case, thus, results in a wave function similar to that in Figure 3-2a, Chapter 3, whose well length now is 2L. When Equation 3-40 is used to calculate the energy of the electron, and $n = 1$,

$$E_S = \frac{n^2h^2}{8mL^2} \rightarrow \frac{1h^2}{8m} \cdot \frac{1}{4L^2} \qquad (3\text{-}40)$$

Thus, when the wells abut, the symmetric case gives an energy which is one fourth that of the widely separated wells (see Equation 5-72).

In the antisymmetric case, the resulting wave function is similar to that shown in Figure 3-2b, Chapter 3. Here the well length is also 2L, but now $n = 2$ in Equation 3-40. This gives

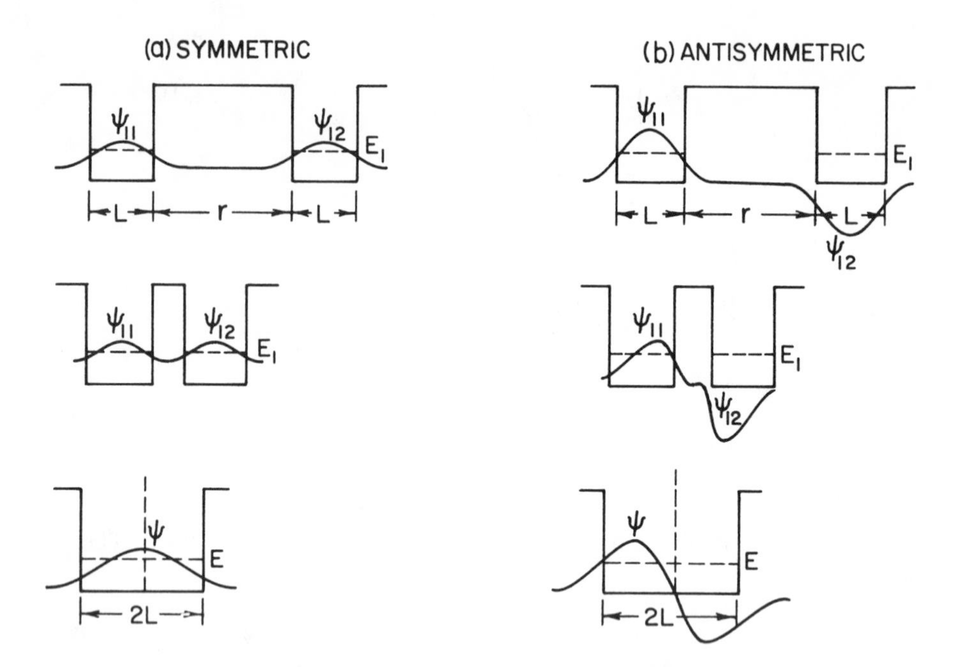

FIGURE 5-14. Changes in wave functions with diminishing well separation. (After Sproull, R. L., *Modern Physics,* John Wiley & Sons, New York, 1956, 206. With permission.)

$$E_A = \frac{n^2h^2}{8mL^2} \rightarrow \frac{4h^2}{8m} \cdot \frac{1}{4L^2} = \frac{h^2}{8mL^2}$$

From this it is seen that the energy of the electron in the antisymmetric case remains unchanged as the wells are brought closer together (Equation 5-72).

The effects of well separation, r, upon the energy of an electron may be shown as in Figure 5-15. This shows that at any finite separation of ions, the ground state (E_1 in this case) splits into two levels. Further, but not derived here, this splitting becomes greater as the distance between them diminishes. It also can be shown that the splitting, or separation, of the levels occurs sooner and more rapidly for the higher energy levels (above the ground state).

The physical significances of the two cases are shown in Figure 5-16. In the symmetric case, a low-energy, stable configuration is one in which the greatest probability density of finding the electron is between the two ions (Figure 5-16a). The stability arises from the +, −, + order which gives rise to coulombic attraction. In the antisymmetric case, the greatest probability of finding an electron (Figure 5-16b) results in the formation of two oppositely charged dipoles. This leads to repulsion and is a higher energy, lower stability configuration than the symmetric case.

One factor omitted thus far is the repulsive forces between the ions. These forces can be estimated from properties of ionic crystals. Here, the repulsive force, which originates from Pauli exclusion, is approximated by

$$W(r) \propto \frac{\beta}{r^s}$$

$$5 < s < 12 \tag{5-73}$$

It is apparent that W(r) is negligible at large distances, but becomes very large at small

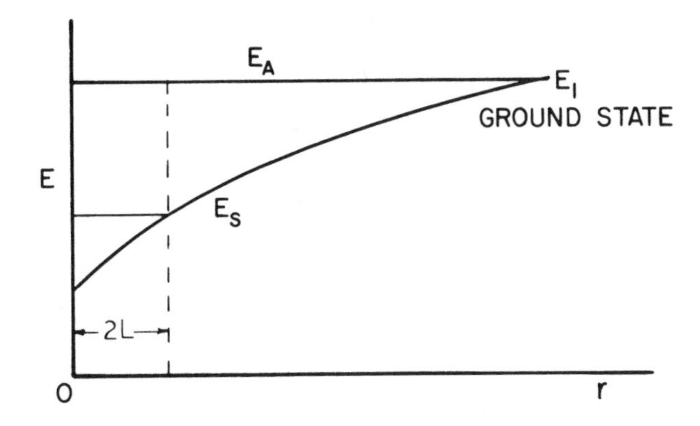

FIGURE 5-15. Effect of well separation upon the energy levels of an electron. (Modified from Sproull, R. L., *Modern Physics*, John Wiley & Sons, New York, 1956, 207. With permission.)

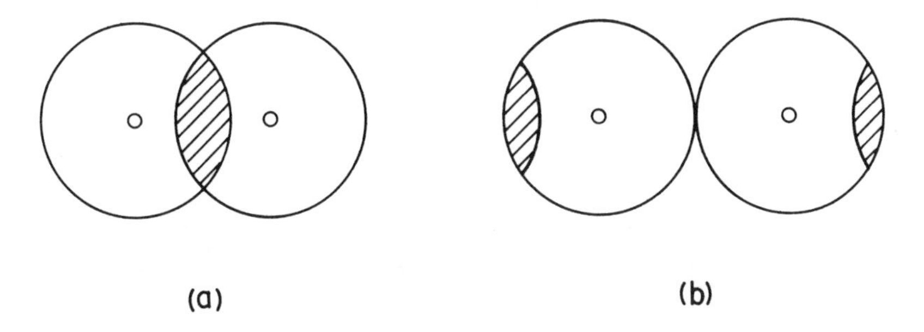

FIGURE 5-16. Schematic representations of the effects of (a) the symmetric and (b) the antisymmetric wave functions.

distances. This must also be included, in terms of the repulsive energy involved, in the model obtained thus far (Figure 5-17). When the attractive energies are added to the repulsive energy, E_R, a minimum occurs in E_S at r_o, the equilibrium interionic distance. The similarity between this curve and that deduced for the thermal expansion of solids (Figure 4-14 in Chapter 4) is apparent. This derivation confirms the curve deduced for the thermal expansion of solids. Here, r_o is the equilibrium interionic distance at 0 K.

Thus far, only the effects of bringing two wells together have been examined. The effects of assembling larger numbers of wells are shown in Figure 5-18. As has been shown, two wells cause two levels. Three wells cause three levels. N wells result in N levels. If N is sufficiently large, as in a real crystal, a quasi-continuum of levels results. The band width increases slowly as N increases, the outer band more than the next inner band.

The inner, completed levels are very tightly bound to their nuclei and are virtually unaffected by the presence of the other ions in the lattice. The outermost shells, whether completely filled or not, provide a basis for the prediction of the properties of solids and for their classification. A band model of a solid is shown schematically in Figure 5-19.

The basis for the classification of solids and the prediction of their properties by the band model depends upon the extent to which the outermost band is filled and whether or not it overlaps other bands. The width of any gaps, or energy intervals, between outer bands also plays a significant part in these predictions.

Consider first normal monovalent metals (Figure 5-20a). By this is meant metals

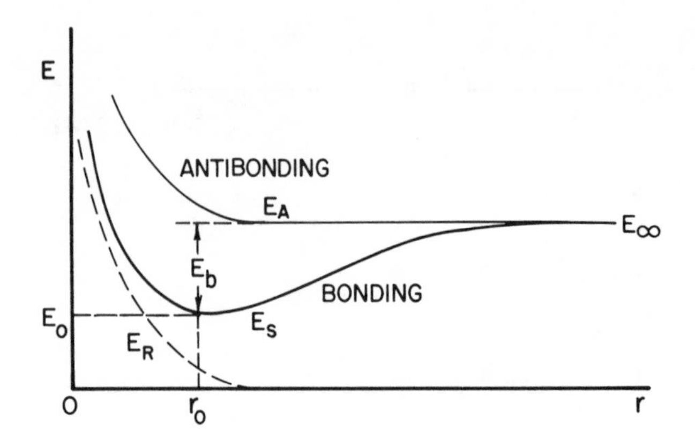

FIGURE 5-17. Attractive and repulsive energies between ions. $|E_b|$ is the minimum energy required to break a bond (dissociate the ions), neglecting zero-point energy.

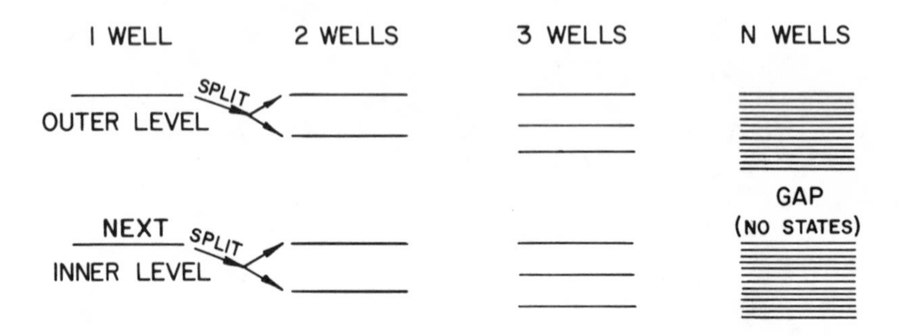

FIGURE 5-18. Effects of assembling numbers of wells.

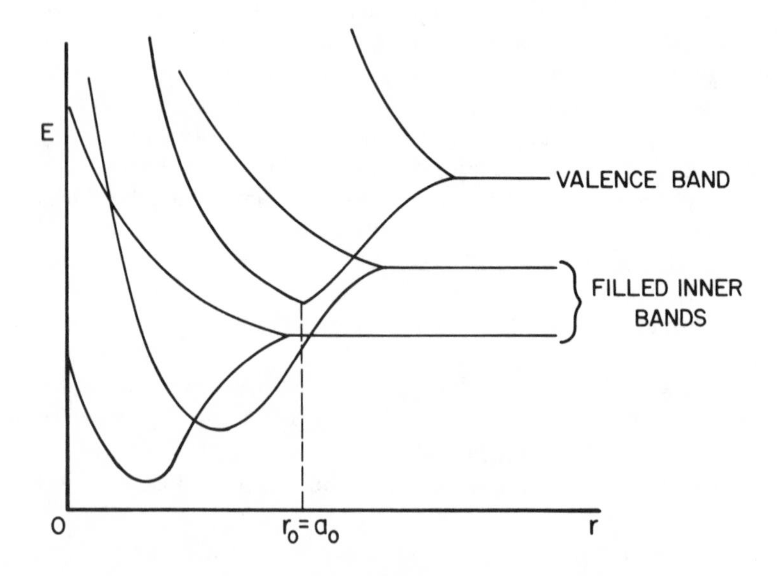

FIGURE 5-19. Schematic band model of a solid. a_o is the equilibrium interionic distance at 0 K.

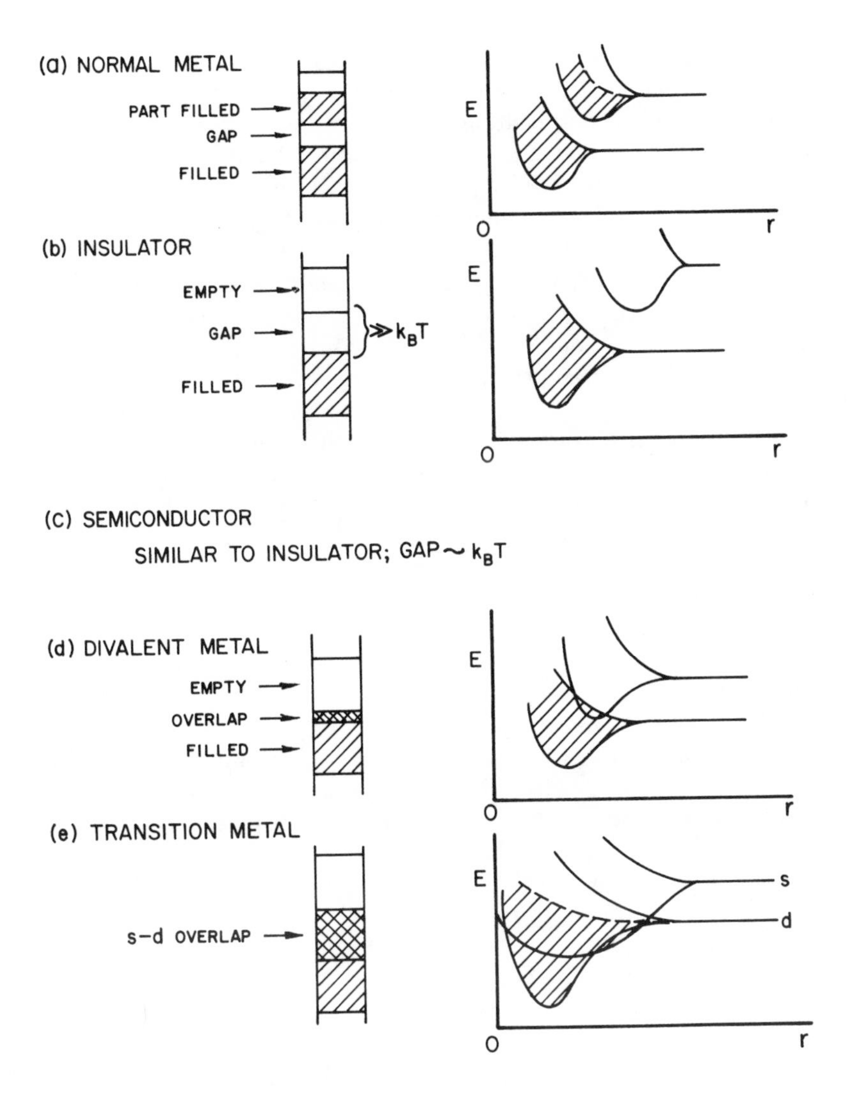

FIGURE 5-20. Band models for the classification of solids.

whose atoms have completed inner bands and only half filled outer, or valence, bands. Since the valence bands are incompletely filled, the application of thermal or electrical energy can excite electrons near E_F to unoccupied levels. They have "some place to go", and can enter into a physical process. Such elements are good thermal and electrical conductors.

Now consider a situation in which the valence bands are completely filled and the next highest band is completely vacant (Figure 5-20b). Much depends upon the extent of the energy gap between them. If the gap is very large ($\gg k_BT$) the element will be an insulator. This results from the fact that ordinary energies cannot cause the valence electrons to participate because no empty states are available for them to occupy and engage in a physical process. Neither thermal nor electrical conduction is possible by the electrons under these conditions. Such elements are insulators. These demonstrate very low ionic thermal and electrical conductivities.

However, if the energy gap between the outermost completely filled valence band and the next outer, completely vacant band (conduction band) is of the order of k_BT, then conduction can occur. Normal energies are sufficient to permit electrons from the top of the filled valence band to jump across the gap and occupy states in the unfilled conduction band. The more electrons that enter into this process, the better the conduction becomes. The resulting "holes" in the formerly filled valence band also can enter into the conduction process. Very pure silicon, germanium, and carbon show this behavior. Elements of this type are known as intrinsic semiconductors. These are discussed in greater detail in Chapter 11.

Some of the divalent elements, such as the alkaline earths (Figure 5-20d), have completed outer s levels. On this basis, it might be thought that such elements should be either insulators or semiconductors. However, these s levels overlap empty p levels. Such elements show conduction behavior similar to that of the normal metals. This is to be expected since the overlap provides the s electrons with easy access to the unoccupied p states.

The transition elements (Figure 5-20f), atoms with incompletely filled outer d and s bands, also show reasonably good metallic behaviors. Here, holes also may enter into the conduction process. The transition elements will be discussed more fully in Chapters 6 and 7, where the effects of alloys are considered.

5.7.1. Applications to the Fermi-Sommerfeld Theory

The band concept may be applied to the Fermi-Sommerfeld theory by considering each band in terms of the individual function for its density of states. This is shown in Figure 5-21. Here the electron behavior is the same as that shown in Figure 5-20, but is in terms of the density of states and energy.

In the case of a normal metal, the outer valence band is only partly filled. Electrons near E_F can accept additional energy and enter into an electron-transport process. The thermal and electrical conductivities of elements such as these would be expected to be good.

The electron configuration of an insulator prevents the electrons from participating in conduction processes. Under normal conditions, the available energy is insufficient to excite outer electrons across the forbidden gap into empty states. Any thermal or electrical conduction, then, must be performed by the ions. Where the energy gap is relatively small, of the order of k_BT, conduction by electrons becomes possible. Normally available energies can promote valence electrons to unoccupied states in the unfilled conduction band. As the energy, or temperature, increases more electrons are promoted. The increased number of activated electrons increases the conductivity and semiconducting behavior is observed.

Divalent metals, such as the alkaline earths, show metallic behavior because of the overlap of the bands. This permits the valence electrons to occupy states of higher energies and to enter into conduction processes without the necessity for jumping across a gap.

The configuration of transition elements, such as Fe, Co, and Ni, appears to consist of two incompletely filled, overlapping, outer bands. This is a very oversimplified model, but is sufficient for present purposes. The electrons share a common Fermi level. Those near E_F, from both bands, enter into physical processes.

5.8. BRILLOUIN ZONE THEORY OF SOLIDS

With the exception of the band theory, the preceding theories consider the electrons to be in a constant potential in the solid. Each ion core, as previously shown in Figure 5-11, constitutes a potential well. In real crystals the ions and, consequently, the potential wells are arranged in periodic arrays. Thus the potential within a crystal is periodic. This periodicity of potential obviously varies with the direction in the crystal. The

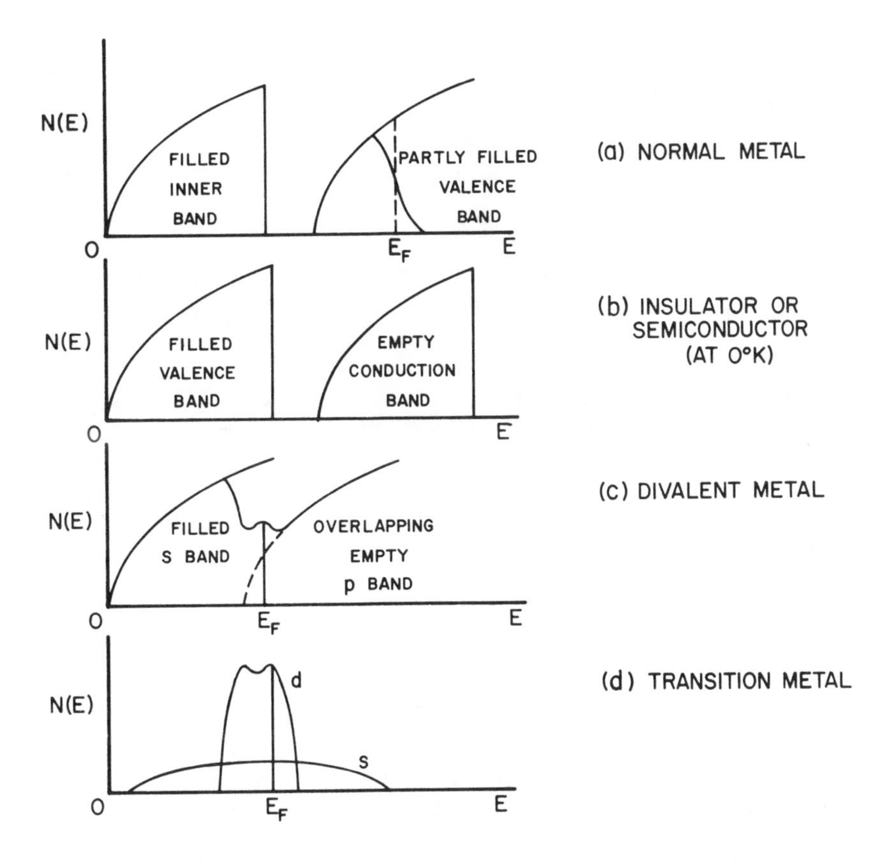

FIGURE 5-21. Combined models for typical classes of elements (figure neglects state variations near tops of bands).

electrons in the crystal interact with the periodic potential. Therefore, the electron transport properties of anisotropic, real crystals must vary with the crystallographic direction. In neglecting the periodic potential, and its variation with lattice direction, the prior theories can deal only with ideal, isotropic materials. Such approaches are best applied to polycrystalline materials; these consist of large numbers of small, randomly oriented grains which average out the anisotropic effects. However, when the directional variation of the internal potential is taken into account, it is possible to account for the anisotropic properties of single crystals.

In the case of the free electron, Equation 3-25,

$$E = \frac{h^2 \bar{k}^2}{8\pi^2 m} \tag{3-25}$$

gives its E-$\bar{k}$ relationship. No restrictions were placed upon $\bar{k}$ so that the energy is a continuous, parabolic function of the wave vector. As will be seen, this must be modified for the case of real crystals.

In the case of real crystals the variation in the magnitude of the periodic potential also varies with the distance between parallel planes of ions. The variation in the potential as measured along the centers of a plane of ions is much greater than that measured half way between two planes (Figure 5-11). The problem of dealing with this behavior can be simplified if most of the electrons are considered to be equidistant from the ionic planes. When this is the case, the variation in the periodic potential from crest to trough is small. This means that there is a nearly equal probability of finding an electron on a potential crest as in a trough. This nearly equal crest-trough probability allows the approximation that the electrons are practically unaffected by the small variation in potential. This is the same as setting $V(x) = 0$ as in Equation 3-22. This permits the approximation that the electrons are "nearly free" and that Equation 3-25 is still applicable.

However, at certain values of $\bar{k} = \bar{k}_c$, the nearly-free model breaks down. Here, even for small potential differences between the crests and troughs, the electrons behave like standing waves for critical values of $\bar{k}$. This arises from the periodicity of the lattice.

Consider electron diffraction similar to that described in Section 2.3, Chapter 2. An electron injected into the lattice "sees" the same periodic potential as an electron present prior to any injection. At certain energies and angles the injected electrons are diffracted from the lattice planes and Bragg's Law is obeyed. Obviously, the electrons already within the lattice and those which are scattered do not satisfy the Bragg relationship.

Consider an incident, plane wave of electrons ($\psi_o = A_o(x)e^{i\bar{k}x}$) to be moving in the x direction perpendicular to a family of lattice planes. As the wave passes across each row of ions, some of it is scattered in waves which radiate equally in all directions from each ion. All such waves from a given plane of ions are in phase since they originated from the same portion of the incident wave. This results in the formation of two sets of plane waves. One of these proceeds in the direction of the incident wave and is in phase with it. The other set of plane waves moves opposite to this ($\psi_1 = A_1(x)e^{-i\bar{k}x}$). Their amplitudes, $A_1(x)$, vary with the magnitude of the periodic potential of the crystal. Since these originate from different planes, destructive interference occurs, or $A_1(x) = 0$. This means that an electron with a suitable $\bar{k}$ will travel undisturbed through the lattice and will behave as a nearly free electron.

Now consider the set of conditions for which Bragg's Law is obeyed. For this case, the Bragg equation gives the conditions for diffraction:

$$n\lambda = 2a \sin\theta \tag{5-74}$$

where a is the interplanar distance and θ is the angle of incidence. From this equation, and the definition of $\bar{k}$,

$$\frac{\bar{k}}{2\pi} = \frac{1}{\lambda} = \frac{n}{2a\sin\theta} \tag{5-75}$$

Constructive interference occurs when twice the interplanar distance is an integral multiple of the wavelength. In the present case, the electrons are moving perpendicular to the plane, $\theta = \pi/2$, and, for this critical condition,

$$\bar{k}_c = \pm\frac{n\pi}{a} \tag{5-76}$$

The condition for Bragg reflection thus represents wave vectors which are forbidden to the valence electrons within the crystal. From Equation 3-25 it can be seen that the disallowed values of $\bar{k}$ result in forbidden energy ranges, or energy gaps. This condition is responsible for the breakdown of the applicability of Equation 3-25, which was previously indicated. The inverse relationship of $\bar{k}$ to a is the basis for the concept of "reciprocal space" when three dimensions are used. It is exactly the same concept employed to describe umklapp processes (Chapter 4).

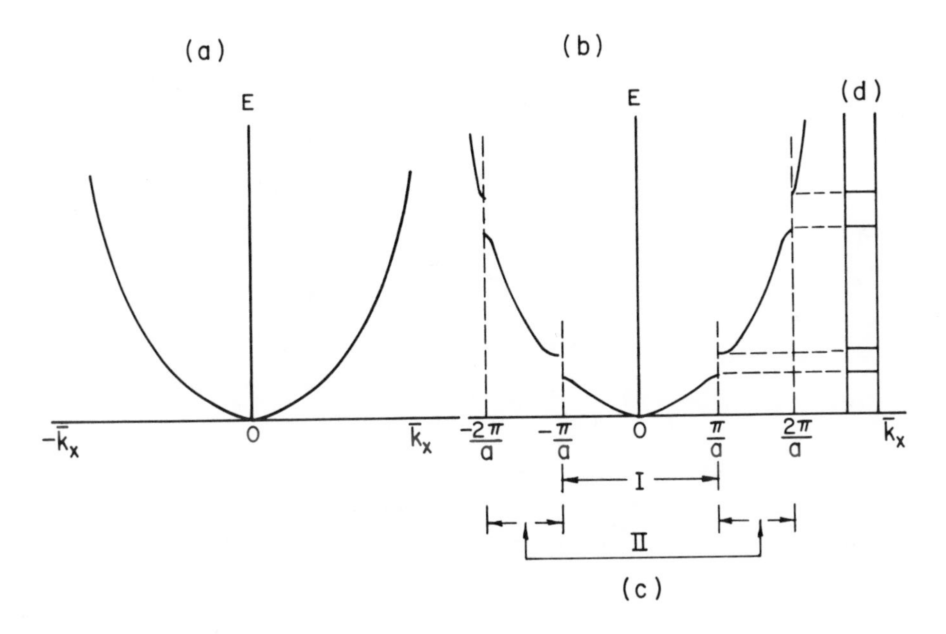

FIGURE 5-22. (a) Free electron theory; (b) electron in a periodic potential; (c) one-dimensional Brillouin zone; (d) band model.

The wavelets which are reflected from the ions of each set of planes are weak (low amplitude). When the conditions are correct nearly all of the incident beam will be reflected.

Thus, at the critical Bragg condition, the resulting incident and reflected waves can be expressed as

$$\psi_I \propto e^{i\bar{k}x} + e^{-i\bar{k}x} \tag{5-77a}$$

which varies as cos $\bar{k}x$ and by

$$\psi_{II} \propto e^{i\bar{k}x} - e^{-i\bar{k}x} \tag{5-77b}$$

which varies as sin $\bar{k}x$. These functions are standing waves. The electrons are travelling back and forth within the lattice. In addition, the electron density is periodic in the lattice. Equation 5-77b has its minima where Equation 5-77a has its maxima. Thus, one of the standing waves has its greatest electron density in the potential troughs between the ions. This means that the energy of such an electron is less than that of a free electron. The other standing wave has its greatest electron density at the crests instead of the potential troughs. Here the energy of an electron is greater than that of a free electron. So, at each critical value of $\bar{k}$, the electron energy divides into two values. The magnitude of the energy separation, or gap, between these values is a function of the magnitude of variation of the lattice potential. Since the magnitude and periodicity of the lattice potential vary with the direction taken in the lattice, it is expected that these energy gaps will occur at different values of $\bar{k}$ and E for different lattice directions. This is the basis for understanding the anisotropic properties of crystals. The surfaces formed in $\bar{k}$ space by the planes perpendicular to the various values of $\bar{k}_c$ enclose a volume which constitutes the Brillouin zone for the given lattice. The one-dimensional case is illustrated in Figure 5-22.

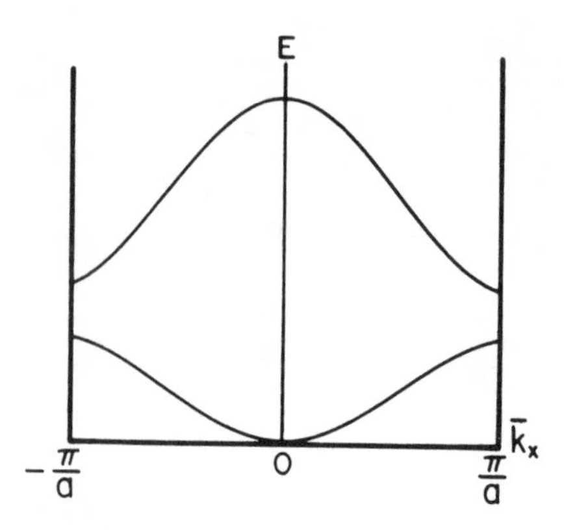

FIGURE 5-23. Reduced zone representation of Figure 5-22b.

The illustration for the case of the free electron, Equation 3-25, shows continuous parabolic behavior, since no restrictions were placed upon $\bar{k}$. The energy discontinuities at the critical values, $\bar{k}_c$, Equation 5-76, result from the limitations placed upon $\bar{k}$ which arise from the periodicity of the lattice. The projection parallel to the $\bar{k}_x$ axis gives the linear Brillouin zone for the x direction. The projection parallel to the E axis gives the band model. The energy gap and the extent to which the bands are filled, as well as any overlap, play exactly the same roles as described in Section 5.7.

Figure 5-22b is sometimes called the "extended zone" model. Here, $\bar{k}_x$ is defined over the range between $\pm\infty$. The same information can be furnished by a "reduced zone" picture in which $\bar{k}_x$ is defined between $\pm\pi/a$. As shown in Figure 5-23, the second and higher zones are "folded" over back into the first zone.

The simplest case for the illustration of the three-dimensional character of the Brillouin zone may be obtained by equating Equations 3-25 and 3-49:

$$E = \frac{h^2\bar{k}^2}{8\pi^2 m} = \frac{h^2}{8m}\left[\frac{n_x^2}{a^2} + \frac{n_y^2}{a^2} + \frac{n_z^2}{a^2}\right] \tag{5-78}$$

in which the interplanar distance, a, is substituted for the well dimension, L. From this it can be seen that

$$\frac{\bar{k}}{\pi} = \pm\frac{1}{a}\left[n_x^2 + n_y^2 + n_z^2\right]^{1/2} \tag{5-79}$$

and

$$\bar{k}_c = \pm\frac{n_i\pi}{a} \tag{5-80}$$

Equations 5-80 and 5-76 are identical.

These concepts will be applied to real lattices to obtain their Brillouin zones.

5.8.1. Geometric Approach

A simple case is presented here as an introduction to the more complex approaches which follow. Consider a two-dimensional lattice as shown in Figure 5-24. From Equations 5-75 and 5-76

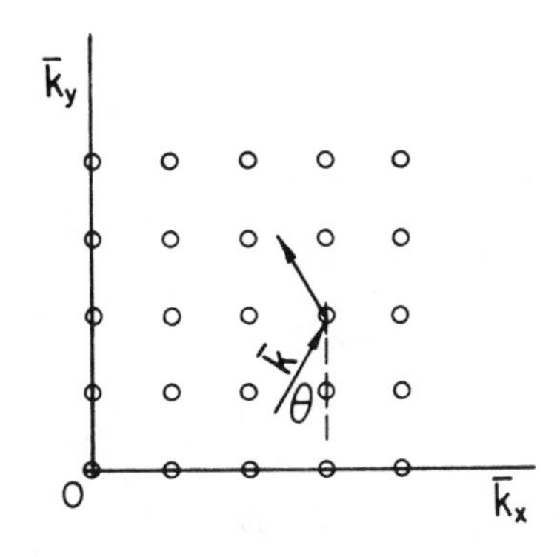

FIGURE 5-24. Diffraction in a two-dimensional lattice in reciprocal space.

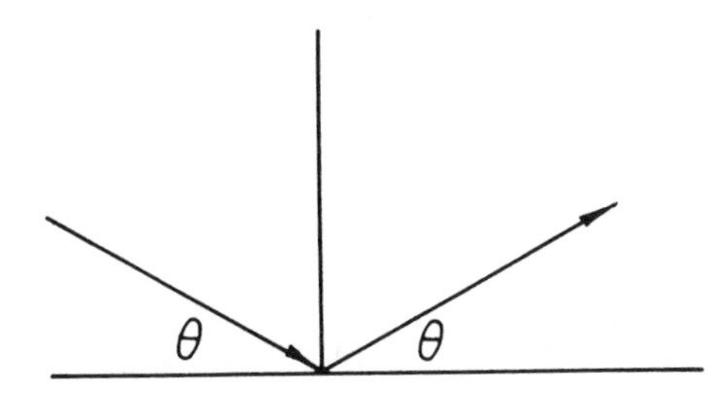

FIGURE 5-25. Bragg reflection from a given plane.

$$\bar{k}_c = \pm \frac{n\pi}{a \sin \theta} = \pm \frac{n\pi}{a} \quad (5\text{-}81)$$

The first discontinuity in the E vs. $\bar{k}$ curve occurs where this condition is first met. This defines the first Brillouin zone. From the figure

$$\bar{k} = \frac{\bar{k}_X}{\sin \theta} \quad (5\text{-}82)$$

Equating Equations 5-81 and 5-82

$$\frac{\bar{k}_X}{\sin \theta} = \pm \frac{n\pi}{a \sin \theta}$$

and the critical value for the X direction again is found to be

$$\bar{k}_{X,c} = \pm \frac{n\pi}{a}$$

The critical value for the Y direction can be found in a similar way. This gives the two-dimensional Brillouin zone for the simple-cubic lattice.

5.8.2. The Ewald Construction and Brillouin Zones

Ewald (1921) devised a method for indexing complex X-ray patterns in terms of the reciprocal lattice. This provided the basis for the determination of Brillouin zones be-

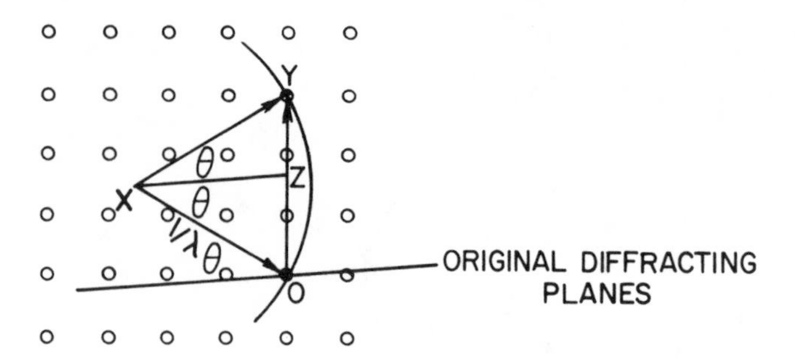

FIGURE 5-26. Vectorial representation of radiation in a reciprocal lattice.

cause the diffracting planes satisfy Bragg's equations and enclose the volume constituting the Brillouin zone.

Consider the incident and reflected waves from any given reflecting planes as shown in Figure 5-25. For first-order reflection, the Bragg condition for reflection can be expressed as

$$\frac{1}{a} = \frac{2}{\lambda} \sin \theta \tag{5-83}$$

This form, where the reciprocal of the interplanar spacing is used, transposes the crystallographic into the reciprocal lattice. The vectorial relationships between $1/a$ and $1/\lambda$ will now be examined.

As shown in Figure 5-26, a vector XO of magnitude $1/\lambda$ which makes the angle θ with the given plane can be obtained from the original set of conditions for the incident beam. This vector is drawn in reciprocal space so that it ends at a lattice point. This lattice point is selected as the origin. The sphere of radius XO, with center at X, will determine all of the possible directions of the reflected rays. The lattice points intersected by the sphere are equivalent reciprocal lattice points. In other words, the points which lie upon the surface of this sphere belong to the same set of planes as that of the original lattice point and correspond to each other. The vector XY, from the center of the sphere to one of these equivalent points, is the equivalent of the original vector XO. The vector YO is perpendicular to the original set of reflecting planes $\{h,k,\ell\}$, since it connects equivalent lattice points in two planes of the same set and equals $1/a$, where a is the distance between the planes. XZ bisects the angle OXY and YO. The magnitude of XO is $1/\lambda$ and $OZ = 1/\lambda \sin \theta$.

Since YO is normal to the set of reflecting planes, and XZ is the perpendicular bisector of YO, XZ must be parallel to the set of reflecting planes. Then YO must be twice the length of OZ; thus $YO = 1/a = 2/\lambda \sin \theta$ satisfies the Bragg equation (Equation 5-83).

The shorter λ is, the larger will be the magnitude of the vector XO. Since XO is the radius of Ewald's sphere, the probability of its intersection with other reciprocal lattice points increases as its magnitude increases. If the sphere did not intersect such a point, the conditions for diffraction would not be fulfilled. The maximum wavelength at which diffraction can occur, as determined from Bragg's law by setting $\theta = \pi/2$, is found to be $\lambda = 2a$. For a given incident wavelength, within this limit, the vector sum XO plus YO must equal XY. Since XO and XY are fixed for a given incident radiation, YO (as previously indicated) will determine when Bragg's law is satisfied.

This can be reexpressed by designating XO as $\bar{k}$, XY as $\bar{k}'$, both equal to $2\pi/\lambda$, and YO as K which is 2π times its original vector in the reciprocal lattice. The multiplication

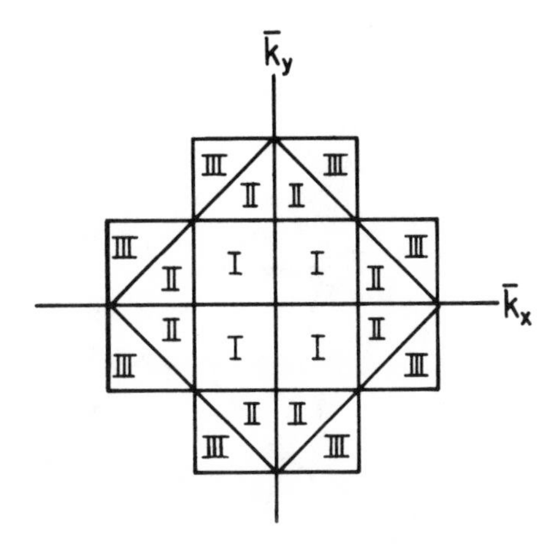

FIGURE 5-27. Brillouin zones for a two-dimensional simple cubic lattice.

of the three vectors involved by 2π maintains $\bar{k}$ and $\bar{k}'$ in terms of the original definition of a wave vector. It also has the effect of expanding the reciprocal lattice. By means of these designations, the Bragg conditions are given in terms of the Ewald geometry as

$$\bar{k} + K = \bar{k}'$$

This may also be expressed as

$$(\bar{k} + K)^2 = \bar{k}^2$$

After expansion and simplification

$$2\bar{k}K + K^2 = 0 \tag{5-84}$$

This gives the condition for Bragg diffraction in terms of K and the wave vector. The vector K may be written, as defined, as

$$K = 2\pi(n_1\bar{b}_1 + n_2\bar{b}_2 + n_3\bar{b}_3) \tag{5-84a}$$

in which n_1, n_2, and n_3 are integers and $\bar{b}_1$, $\bar{b}_2$, and $\bar{b}_3$ are the unit vectors in the reciprocal lattice. The wave vector satisfying Equation 5-84 then must be

$$\bar{k} = \pi(n_1\bar{b}_1 + n_2\bar{b}_2 + n_3\bar{b}_3) \tag{5-84b}$$

Equation 5-84b is used generally in the determination of the Brillouin zones of crystal lattices. In two dimensions, the lines perpendicular to the appropriate vectors enclosed an area which constitutes the Brillouin zone. Correspondingly, in three dimensions, the planes perpendicular to these vectors enclose the volume which is the Brillouin zone. These are the reflecting planes at which Bragg's law is obeyed. Those with smaller $\bar{k}$ (longer λ) cannot have $\bar{k}'$ which will intersect with equivalent points in the reciprocal lattice (Equation 5-84) is not obeyed.

The simple-cubic lattice is used to provide a two-dimensional illustration. The reciprocal lattice vectors are given in terms of the lattice parameter, a, as

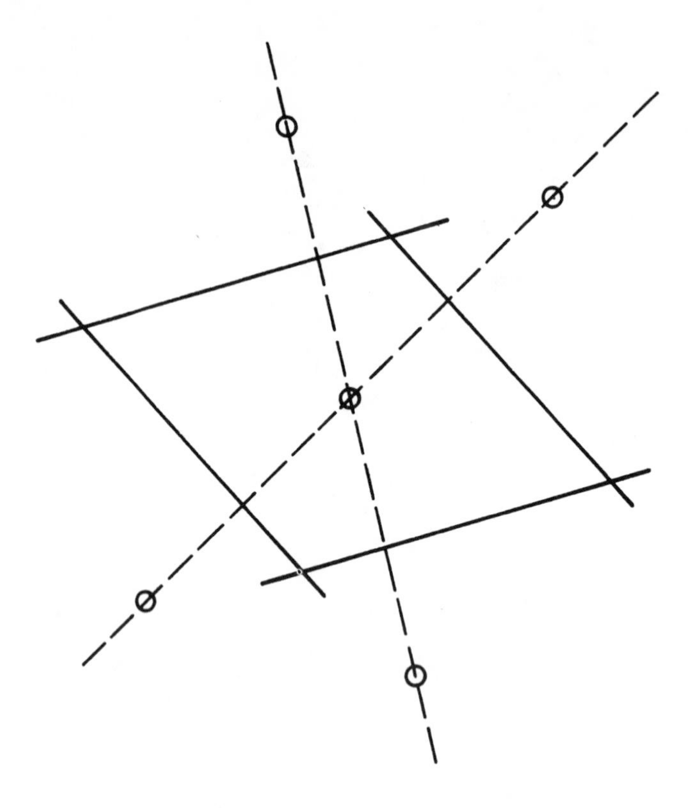

FIGURE 5-28. Construction of a two-dimensional proximity cell in reciprocal space.

$$\bar{b}_1 = \bar{b}_2 = 1/a$$

$$\bar{b}_3 = 0$$

In this two-dimensional case, combinations of n_1 and n_2 equal to zero and ±1 give the first Brillouin zone. The second zone is formed by the minimum, additional area, beyond that of the first zone, for combinations of n_1 and n_2 set at ±1, as shown in Figure 5-27.

Once the foregoing is understood, Brillouin zones may be constructed simply by connecting nearest neighbors in the reciprocal lattice by reciprocal lattice vectors. Planes are then passed perpendicularly through the midpoints of the connecting vectors. The minimum volume enclosed by these planes constitutes the first Brillouin zone. The second zone is constructed by planes bisecting the vectors to next-nearest neighbors which enclose a minimum volume. This follows directly from the knowledge of the Bravais lattice of the crystal being considered. However, such a construction, by itself, provides no information regarding the nature of the electron behavior within the zone. These have been called "proximity cells." Such a construction is shown in Figure 5-28 for two dimensions.

This approach, originated by Wigner and Seitz (1933, 1934), has been used to approximate spherically symmetric forms of Schrödinger's equation in real crystal space. These simplified the calculations for the elastic constants and bonding energies of crystals. The results were in reasonable agreement with the observed data.

Brillouin zones possess interesting and important properties. Some of these are apparent from Figure 5-27, for two dimensions. All zones have equal volume. Each zone must abut on a previous zone and must be capable of translation to either a higher or a lower zone. This is apparent in the figure by simple "folding" of one zone, or a part of a zone, into another. It is the basis for the reduced zone representation shown in Figure 5-23.

The surfaces enclosing the zone correspond to the critical Bragg conditions. Lattices with the same symmetry will have the same Brillouin zones, except for dimensional variations. This is a result of the fact that no two elements have the same lattice parameters. Each zone can accommodate two electron states per ion when spin is taken into account. A simple way of verifying this is to note that Equation 5-79 is based upon Equation 3-49; this equation implicitly includes two electron spins for each state.

Another way to calculate the number of electron states which can be accommodated in a zone is to consider the number of states which an ion can accommodate. This will be approached by considering the least complicated case, that of a simple-cubic lattice. The distance, d, between any two crystal planes of the same {h,k,l} is

$$d = Na \tag{5-85}$$

where a is the interionic spacing and N is the number of interplanar distances separating two planes of the given set. The factor N also represents the number of ions which lie on the normal to both planes when N is relatively large.

The wave functions for the electrons will be periodic functions of the interionic spacing. These will have an integral number of half-wavelengths between the two planes of interest. Such behavior is given by $\lambda = 2d/n$ or $1/\lambda = n/2d$. The substitution of Equation 5-85 into this gives the wave number as

$$\frac{1}{\lambda} = \frac{n}{2d} = \frac{n}{2Na} \tag{5-86}$$

The wave number will be doubled when spin is included. An expression for the wave vectors can be obtained from this factor included in Equation 5-86 as

$$\bar{k} = \frac{2\pi}{\lambda} = 2\pi \cdot \frac{2n}{2Na} = \frac{2\pi n}{Na} \tag{5-87}$$

For the first zone (Equation 5-76) $\bar{k}_c = \bar{k}_{max} = \pi/a$. Equating this with Equation 5-87 results in

$$\frac{2\pi n}{Na} = \frac{\pi}{a}$$

$$\frac{2n}{N} = 1$$

$$2n = N \tag{5-88}$$

Thus, when spin is taken into account, two states per ion can be accommodated in the zone. From this it is apparent that the Brillouin zone for an element with an odd number of valence electrons must be half filled. Those with even numbers will be completely filled. The properties of the latter will be determined, as in band theory, by the magnitude of the energy gap or any overlap between successive zones. Because of this, only one, or at most two, zones are needed to characterize the properties of the solid. This will be discussed more fully in the last section of this chapter.

The Brillouin zones of greatest interest are shown in Figure 5-29.

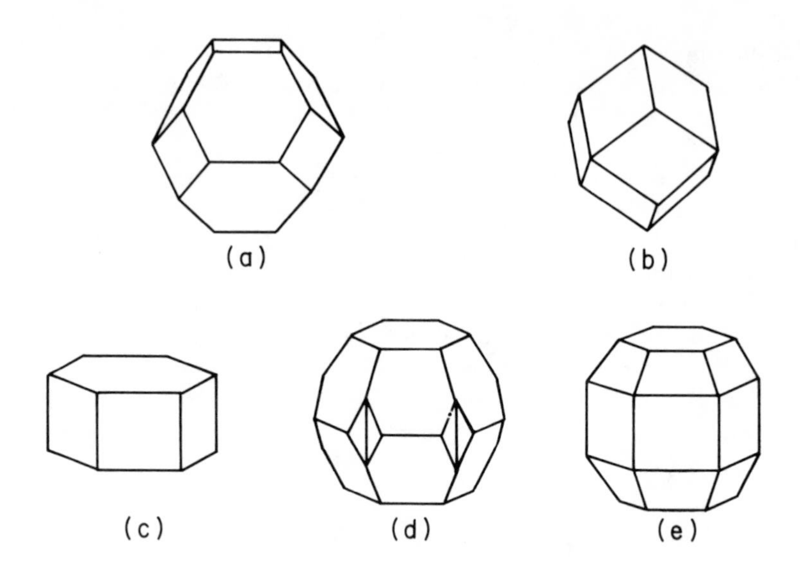

FIGURE 5-29. Common Brillouin zones. (a) FCC; (b)BCC; (c) HCP first zone; (d) HCP second zone; (e) Jones zone.

It is important to note that Equation 5-88 does not hold for the Jones zone for the hexagonal close packed (HCP) lattice because it is not based upon a permissible Bravais lattice. This is discussed more fully in the vector treatment of hexagonal lattice in Section 5.8.3.4. The number of states which can be accommodated by this Brillouin zone for HCP elements is determined by the axial, or c/a, ratio. This is given approximately by

$$n = 2 - \frac{3}{4}\left[\frac{a}{c}\right]^2\left\{1 - \frac{1}{4}\left[\frac{a}{c}\right]^2\right\} \qquad (5\text{-}88a)$$

For the ideal case of the closest packing of hard spheres, $c/a = (8/3)^{1/2} = 1.633$ and $n = 1.745$ states per ion. In the case of zinc, for example, $c/a = 1.856$ and $n = 1.792$. This is significantly different from the two states per ion which can be accommodated by the zones of the FCC and BCC systems. It is helpful in determining alloy phase relationships as shown in Chapter 10.

Electrons within a given Brillouin zone may be treated as being nearly free when the Fermi surface is unaffected by the zone boundary. That is, when the wave vector at the Fermi surface is less than $\bar{k}_c$. The energy of the electron will lie upon the parabolic portion of the E vs $\bar{k}$ relationship given by Equation 3-25 when this is the case. This approximation does not hold for electrons with wave vectors, or energies, close to the zone boundaries. Here a considerable increase in $\bar{k}$ occurs near $\bar{k}_c$ for a relatively small change in energy (Figure 5-22); the curve no longer is parabolic and electrons show different behaviors.

An insight into these differences may be obtained by an examination of the implications of Equation 3-25. The second derivative of this equation for a free electron is a constant given by

$$\frac{d^2E}{d\bar{k}^2} = \frac{h^2}{4\pi^2 m} = \frac{\hbar^2}{m}$$

Here, the electron mass is constant and equal to that of a free electron. Departures from "free" behavior are visualized more readily by rearranging the above equation to read

$$m^* = \hbar^2 \left[\frac{d^2E}{d\bar{k}^2}\right]^{-1}$$

Electrons with wave vectors very close to $\bar{k}_c$ do not obey the parabolic relationship (Figure 5-22). The "curvature," as given by the second derivative, is changed significantly. This results in a change in the "effective mass" of such electrons. This parameter is designated by m*.

In the cases of normal univalent and trivalent metals, where the zones are partially filled, the electrons may be approximated to be nearly free and to have effective masses close to those of free electrons. The same is true for divalent metals because of overlapping zones. All of the equations given in Section 5.6 which include electron mass should be considered as involving m*. This also was noted earlier for Equations 5-21 and 5-61 (see Table 11.1, in Chapter 11).

The role of the effective mass of an electron is very important in explanations of the properties of semiconductors. The properties of electrons very close to zone boundaries are derived in a more rigorous way and discussed in detail in Chapter 11, on semiconductors, where they are of greatest applicability.

5.8.3. Vector Treatment of Reciprocal Lattices

It was shown that planes perpendicular to the vectors (Equation 5-48b) derived from Ewald's treatment of diffraction in reciprocal lattices, enclose the Brillouin zones. All that is required to construct the Brillouin zone for a given unit cell is that the reciprocal lattice vectors be known.

The properties and mathematics governing vectors and their manipulation can be found in most introductory physics and mathematics texts. The few relationships needed in this section are given here for convenience because the "mechanics" of the construction of Brillouin zones require the use of vectors. Let $\bar{a}_1$, $\bar{a}_2$, and $\bar{a}_3$ be real lattice vectors for the three orthogonal directions, and $\bar{b}_1$, $\bar{b}_2$, and $\bar{b}_3$ be their corresponding vectors in the reciprocal lattice. By definition

$$\bar{a}_u \cdot \bar{b}_v = \begin{vmatrix} 1, u = v \\ \\ 0, u \neq v \end{vmatrix}$$

In other words,

$$\bar{a}_1 \cdot \bar{b}_1 = \bar{a}_2 \cdot \bar{b}_2 = \bar{a}_3 \cdot \bar{b}_3 = 1$$

and

$$\bar{a}_1 \cdot \bar{b}_2 = \bar{a}_1 \cdot \bar{b}_3 = \bar{a}_2 \cdot \bar{b}_3 = \bar{a}_3 \cdot \bar{b}_1 = \bar{a}_3 \cdot \bar{b}_2 = 0$$

The reciprocal lattice vectors are given in terms of the real lattice vectors by

$$b_1 = \frac{\bar{a}_2 \times \bar{a}_3}{\bar{a}_1 [\bar{a}_2 \times \bar{a}_3]} = \frac{\bar{a}_2 \times \bar{a}_3}{V} \qquad \text{(5-89a)}$$

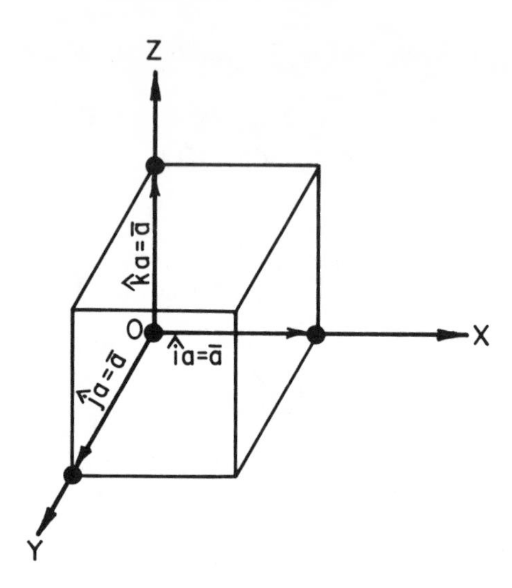

FIGURE 5-30. Vectors for the simple cubic lattice.

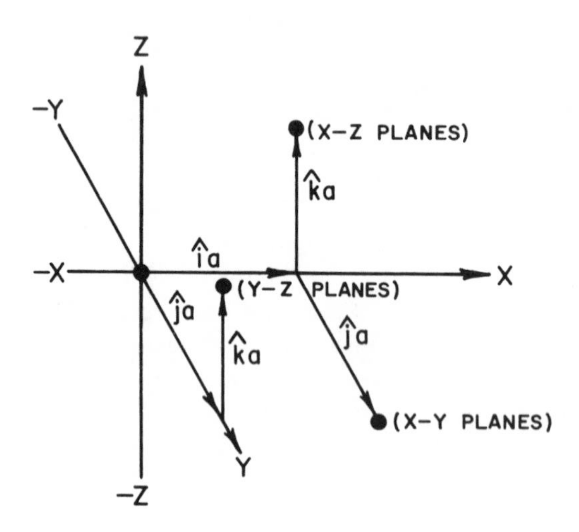

FIGURE 5-31. Vectors for the face-centered cubic lattice.

where V is the volume of the unit cell and

$$\bar{b}_2 = \frac{\bar{a}_3 \times \bar{a}_1}{V} \tag{5-89b}$$

$$\bar{b}_3 = \frac{\bar{a}_1 \times \bar{a}_2}{V} \tag{5-89c}$$

These vectors will define the planes which constitute the boundaries of the Brillouin zones for any properly selected unit cell.

5.8.3.1. The Simple Cubic Lattice

The primitive translations, that is, those which transpose, or map, a given point into an equivalent point in the real lattice, are given by

$$\bar{a}_1 = a\hat{i}$$

$$\bar{a}_2 = a\hat{j}$$

$$\bar{a}_3 = a\hat{k}$$

where $\hat{i}$, $\hat{j}$ and $\hat{k}$ are the unit vectors and the lattice parameter is given by a. The unit cell is shown in Figure 5-30. By means of Equation 5-89 the reciprocal lattice vectors are:

$$\bar{b}_1 = \frac{\bar{a}_2 \times \bar{a}_3}{V} = \frac{a^2 \begin{vmatrix} \hat{i} & \hat{j} & \hat{k} \\ 0 & 1 & 0 \\ 0 & 0 & 1 \end{vmatrix}}{a^3} = \frac{\hat{i}}{a}$$

$$\bar{b}_2 = \frac{\bar{a}_3 \times \bar{a}_1}{a^3} = \frac{a^2 \begin{vmatrix} \hat{i} & \hat{j} & \hat{k} \\ 0 & 0 & 1 \\ 1 & 0 & 0 \end{vmatrix}}{a^3} = \frac{\hat{j}}{a}$$

$$\bar{b}_3 = \frac{\bar{a}_1 \times \bar{a}_2}{a^3} = \frac{a^2 \begin{vmatrix} \hat{i} & \hat{j} & \hat{k} \\ 1 & 0 & 0 \\ 0 & 1 & 0 \end{vmatrix}}{a^3} = \frac{\hat{k}}{a}$$

These now may be included in Equation 5-84b and result in:

$$\bar{k} = \pi(n_1\bar{b}_1 + n_2\bar{b}_2 + n_3\bar{b}_3) = \frac{n_1\pi}{a}\begin{vmatrix} 1 \\ 0 \\ 0 \end{vmatrix} + \frac{n_2\pi}{a}\begin{vmatrix} 0 \\ 1 \\ 0 \end{vmatrix} + \frac{n_3\pi}{a}\begin{vmatrix} 0 \\ 0 \\ 1 \end{vmatrix} \quad (5\text{-}90)$$

This gives the general equation for the vectors perpendicular to the planes of the Brillouin zones of the simple cubic structure. The substitution of suitable values for n_1, n_2, and n_3, as indicated in Section 5.8.2, give the vectors for the various zones shown in Figure 5-27. In three dimensions, the reciprocal lattice is a cube; this lattice is its own reciprocal.

5.8.3.2. Face-Centered Cubic Lattice

The primitive translations are taken as the vectors from a corner ion to each of one of three nearest face ions lying in the three planes which intersect at the corner ion. This is shown in Figure 5-31. Here a is taken as *one half* the lattice parameter for simplification. The primitive translations are:

$$\bar{a}_1 = \hat{i}a + \hat{j}a = a(\hat{i} + \hat{j})$$

$$\bar{a}_2 = \hat{i}a + \hat{k}a = a(\hat{i} + \hat{k})$$

$$\bar{a}_3 = \hat{j}a + \hat{k}a = a(\hat{j} + \hat{k})$$

The numerator of Equation 5-89a is

$$\bar{a}_2 \times \bar{a}_3 = a^2 \begin{vmatrix} \hat{i} & \hat{j} & \hat{k} \\ 1 & 0 & 1 \\ 0 & 1 & 1 \end{vmatrix} = a^2(-\hat{i} - \hat{j} + \hat{k})$$

so the denominators of Equation 5-89 will be

$$V = \bar{a}_1 \cdot [\bar{a}_2 \times \bar{a}_3] = a^3(\hat{i} + \hat{j})(-\hat{i} - \hat{j} + \hat{k}) = \left| a^3(-\hat{i}\hat{i} - \hat{j}\hat{j}) \right|$$

$$V = 2a_3$$

From Equation 5-89 the reciprocal lattice vectors are:

$$\bar{b}_1 = \frac{\bar{a}_2 \times \bar{a}_3}{2a^3} = \frac{a^2(-\hat{i} - \hat{j} + \hat{k})}{2a^3} = \frac{-\hat{i} - \hat{j} + \hat{k}}{2a}$$

$$\bar{b}_2 = \frac{\bar{a}_3 \times \bar{a}_1}{2a^3} = \frac{a^2 \begin{vmatrix} \hat{i} & \hat{j} & \hat{k} \\ 0 & 1 & 1 \\ 1 & 1 & 0 \end{vmatrix}}{2a^3} = \frac{-\hat{i} + \hat{j} - \hat{k}}{2a}$$

$$\bar{b}_3 = \frac{\bar{a}_1 \times \bar{a}_2}{2a^3} = \frac{a^2 \begin{vmatrix} \hat{i} & \hat{j} & \hat{k} \\ 1 & 1 & 0 \\ 1 & 0 & 1 \end{vmatrix}}{2a^3} = \frac{\hat{i} - \hat{j} - \hat{k}}{2a}$$

Thus, from Equation 5-84b the general form for the vectors perpendicular to the planes of the Brillouin zones of the face-centered cubic structure is given by the vectors

$$\bar{k} = \frac{n_1 \pi}{2a} \begin{vmatrix} -1 \\ -1 \\ +1 \end{vmatrix} + \frac{n_2 \pi}{2a} \begin{vmatrix} -1 \\ +1 \\ -1 \end{vmatrix} + \frac{n_3 \pi}{2a} \begin{vmatrix} +1 \\ -1 \\ -1 \end{vmatrix} \qquad (5\text{-}91)$$

Again, the substitution of suitable values for n_1, n_2, and n_3 will give the vectors perpendicular to the planes which constitute the first Brillouin zone shown in Figure 5-29.

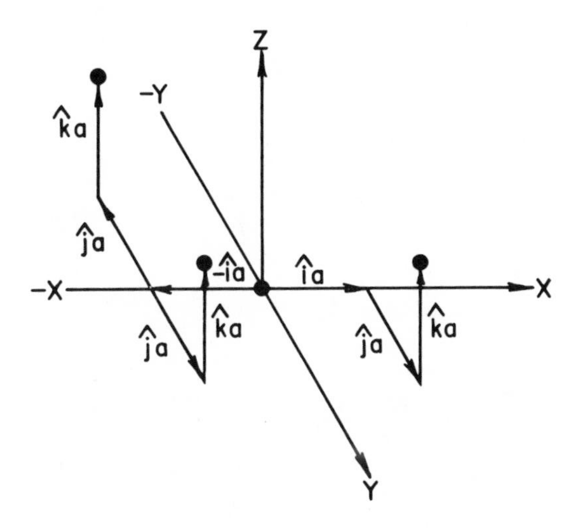

FIGURE 5-32. Vectors for the body-centered cubic lattice.

5.8.3.3. Body-Centered Cubic Lattice

The primitive translations are taken as the vectors from a body ion to each of three corner ions. Again, a is taken as *one half* of the lattice parameter (Figure 5-32).

The primitive translations are:

$$\bar{a}_1 = a(\hat{i} + \hat{j} + \hat{k})$$

$$\bar{a}_2 = a(-\hat{i} + \hat{j} + \hat{k})$$

$$\bar{a}_3 = a(-\hat{i} - \hat{j} + \hat{k})$$

The numerator of Equation 5-89a is

$$a_2 \times a_3 = a^2 \begin{vmatrix} \hat{i} & \hat{j} & \hat{k} \\ -1 & 1 & 1 \\ -1 & -1 & 1 \end{vmatrix} = 2a^2(\hat{i} + \hat{k})$$

and its denominator is

$$\bar{a}_1 \cdot [\bar{a}_2 \times \bar{a}_3] = a(\hat{i} + \hat{j} + \hat{k})2a^2(\hat{i} + \hat{k}) = 4a^3$$

Then, the reciprocal lattice vectors are:

$$\bar{b}_1 = \frac{2a^2(\hat{i} + \hat{k})}{4a^3} = \frac{\hat{i} + \hat{k}}{2a}$$

also

$$\bar{b}_2 = \frac{\bar{a}_3 \times \bar{a}_1}{4a^3} = \frac{a^2 \begin{vmatrix} \hat{i} & \hat{j} & \hat{k} \\ -1 & -1 & 1 \\ 1 & 1 & 1 \end{vmatrix}}{4a^3} = \frac{-\hat{i} + \hat{j}}{2a}$$

and

$$\bar{b}_3 = \frac{\bar{a}_1 \times \bar{a}_2}{4a^3} = \frac{a^2 \begin{vmatrix} \hat{i} & \hat{j} & \hat{k} \\ 1 & 1 & 1 \\ -1 & 1 & 1 \end{vmatrix}}{4a^3} = \frac{-\hat{j} + \hat{k}}{2a}$$

The general equation for the vectors perpendicular to the planes of the Brillouin zones of the body-centered cubic lattice is, from Equation 5-84b,

$$\bar{k} = \frac{n_1 \pi}{2a} \begin{vmatrix} 1 \\ 0 \\ 1 \end{vmatrix} + \frac{n_2 \pi}{2a} \begin{vmatrix} -1 \\ 1 \\ 0 \end{vmatrix} + \frac{n_3 \pi}{2a} \begin{vmatrix} 0 \\ -1 \\ 1 \end{vmatrix} \qquad (5\text{-}92)$$

These vectors also will define the planes enclosing the first zone as shown in Figure 5-29.

5.8.3.4. Hexagonal Lattice

The primitive translations of the hexagonal lattice are based upon the central ion of the basal plane. In contrast to the treatments given earlier for the two common cubic lattices, the lattice parameter, a, is used as the unit of measurement in the basal plane. As is customary, c is the distance between successive basal planes (Figure 5-33).

It must be noted that the hexagonal structure is not a primitive Bravais lattice; this is a right, rhombic prism. In dealing with the HCP structure, it is customary to treat it in terms of *one* of the three right, rhombic prisms of which it is composed. A given lattice point in this HCP unit cell cannot be translated into an equivalent point in an adjacent unit cell by the simple substitution of the vector from the origin to a given lattice point by a corresponding vector to an equivalent point in an adjacent cell. Therefore, the HCP unit cell is not a Bravais lattice, while that of the hexagonal lattice is.

The primitive translations of the hexagonal lattice are:

$$\bar{a}_1 = c\hat{k}$$

$$\bar{a}_2 = \frac{a}{2}\hat{i} + \frac{\sqrt{3}}{2} a\hat{j}$$

$$\bar{a}_3 = \frac{a}{2}\hat{i} + \frac{\sqrt{3}}{2} a\hat{j}$$

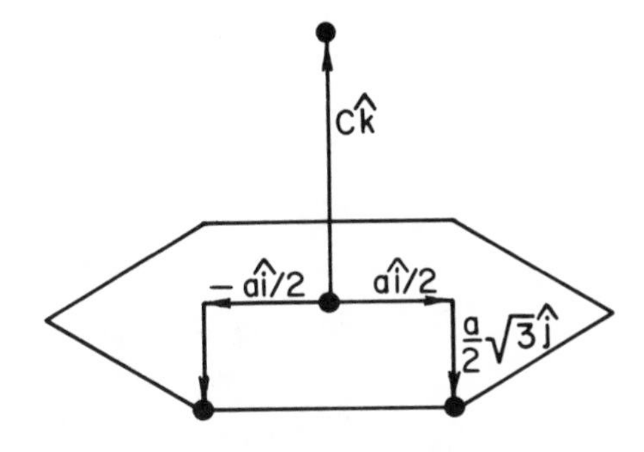

FIGURE 5-33. Vectors for the hexagonal lattice.

The numerator of Equation 5-89a is

$$\bar{a}_2 \times \bar{a}_3 = \begin{vmatrix} \hat{i} & \hat{j} & \hat{k} \\ \frac{a}{2} & \frac{a\sqrt{3}}{2} & 0 \\ \frac{a}{2} & \frac{a\sqrt{3}}{2} & 0 \end{vmatrix} = \frac{\sqrt{3}}{2} a^2 \hat{k}$$

The denominator is

$$\bar{a}_1 \cdot (\bar{a}_2 \times \bar{a}_3) = c\hat{k} \cdot \frac{\sqrt{3}}{2} a^2 \hat{k} = \frac{\sqrt{3}}{2} a^2 c$$

From these

$$\bar{b}_1 = \frac{\frac{\sqrt{3}}{2} a^2 \hat{k}}{\frac{\sqrt{3}}{2} a^2 c} = \frac{\hat{k}}{c}$$

The other reciprocal lattice vectors are:

$$\bar{b}_2 = \frac{\bar{a}_3 \times \bar{a}_1}{\frac{\sqrt{3}}{2} a^2 c} = \frac{\begin{vmatrix} \hat{i} & \hat{j} & \hat{k} \\ -\frac{a}{2} & \frac{a\sqrt{3}}{2} & 0 \\ 0 & 0 & c \end{vmatrix}}{\frac{\sqrt{3}}{2} a^2 c} = \frac{\hat{i}}{a} + \frac{\hat{j}}{\sqrt{3}\, a}$$

and

$$\bar{b}_3 = \frac{\bar{a}_1 \times \bar{a}_2}{\frac{\sqrt{3}}{2} a^2 c} = \frac{\begin{vmatrix} \hat{i} & \hat{j} & \hat{k} \\ 0 & 0 & c \\ \frac{a}{2} & \frac{a\sqrt{3}}{2} & 0 \end{vmatrix}}{\frac{\sqrt{3}}{2} a^2 c} = -\frac{\hat{i}}{a} + \frac{\hat{j}}{\sqrt{3}\,a}$$

From these the general equation for the vectors for the planes of the Brillouin zones of the hexagonal lattice is, from Equation 5-84b,

$$\bar{k} = \frac{n_1 \pi}{c} \begin{vmatrix} 1 \\ 0 \\ 0 \end{vmatrix} + \frac{n_2 \pi}{a} \begin{vmatrix} 1 \\ \frac{1}{\sqrt{3}} \\ 0 \end{vmatrix} + \frac{n_3 \pi}{a} \begin{vmatrix} -1 \\ \frac{1}{\sqrt{3}} \\ 0 \end{vmatrix} \tag{5-93}$$

As is the case for the simple cubic lattice, examination shows that the vector products $\bar{a}_1 \cdot \bar{b}_1 = \bar{a}_2 \cdot \bar{b}_2 = \bar{a}_3 \cdot \bar{b}_3 = 1$; the hexagonal crystal lattice remains unchanged by the transformation. It is hexagonal in reciprocal space. Since the primitive cell ($1/8 \times 8$ shared corner ions = one ion per unit cell) and will accommodate two electron states in the first zone.

The first Brillouin zone of the HCP lattice (Figure 5-29) is also a hexagonal prism. Its basis is not a primitive unit cell because of the ion at the c/2 position. The hexagonal faces of the first zone do not always constitute a barrier for electrons. This arises from the fact that even though this zone is filled, as shown below, overlap can occur across these zone boundaries. Since no energy gap exists at these faces, the second zone must be employed to provide the energy discontinuities.

The number of states in the first Brillouin zone of the HCP lattice is computed as follows: $1/8 \times 8$ shared corner ions at the corners of the unit cell plus the ion at c/2 gives two ions for each Brillouin zone. Metals which crystallize in the HCP lattice usually have a valence of two. Consequently, in the absence of overlap, the first zone would be filled. The second zone also can accommodate two states per ion. Conduction occurs in these metals as a result of the overlap of the filled first zone into the empty second zone. The smallest volume in reciprocal space which was thought to provide energy discontinuities at all zone faces was devised by H. Jones (1934). It consists of a hybrid combination of both first and second zones (see Figure 5-29). The prism faces of the first zone constitute the central surfaces of this combination. The remainder of the combined zone (top and bottom) consists of those surfaces of the second zone which are not replaced by the central prism faces from the first zone. This configuration accounts for the electron:ion ratio being less than two, as given by Equation 5-88a, since its volume is smaller than that of the original second zone.

It is considered that the zone overlap at the hexagonal boundary faces of the Jones zone is most probable for metals with axial ratios more than that of the ideal. Such overlap is not expected when the c/a ratio is smaller than 1.633. These differences have been used to help explain variations in the electrical and magnetic properties of these two classes of HCP elements and their alloys.

Overlap at another set of Jones zone faces can occur when the c/a ratio is less than the ideal. In this case the central, rectangular faces do not represent energy discontin-

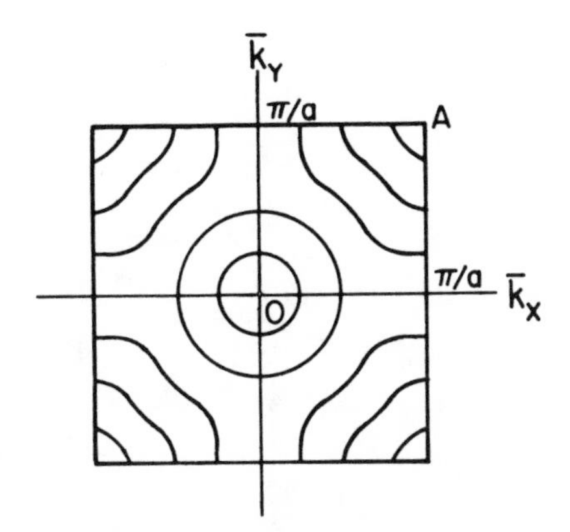

FIGURE 5-34. Isoenergy contours in a two-dimensional Brillouin zone.

uities and overlap may take place. In addition it should be noted that the intersections of these central, rectangular faces with all other adjacent surfaces do not constitute energy discontinuities for either class of c/a ratio.

5.8.4. Physical Significance

As in band theory, the ways in which the electrons are accommodated by the Brillouin zones determine the properties of the solid. This permits the classification of solids and the prediction of their properties as will be shown below. The Brillouin zone clearly shows the effect of crystalline anisotropy on the energies of the electrons, and consequently upon physical properties.

The nearly free electrons can be treated by Equation 3-25 and 3-49 as indicated by Equation 5-78

$$E = \frac{h^2 \bar{k}^2}{8\pi^2 m} = \frac{h^2}{8m}\left[\frac{n_x^2}{a^2} + \frac{n_y^2}{a^2} + \frac{n_z^2}{a^2}\right] \qquad (5\text{-}78)$$

Then, for given energies, isoenergy contours can be drawn in $\bar{k}$ space. These will be nearly spherical surfaces for a cubic lattice, since E varies as $\bar{k}^2$. A breakdown of this behavior occurs when the electron energies approach critical values of $\bar{k} = \bar{k}_c$ (Equation 5-76 and Figure 5-22b). Here the isoenergy surface contours are no longer spherical.

This is shown in Figure 5-34, in two dimensions.

As the electron density, or the Fermi surface, approaches the zone wall, the values of $\bar{k}$ become increasingly greater than that predicted by the nearly free electron theory. This continues until the discontinuity at the zone wall, $\bar{k}_c$, is encountered. When the entire zone is filled, it accommodates an electron:ion ratio of two (Equation 5-88).

It can be seen that the energy-wave vector relationships along the [100], or $\bar{k}_x$: direction will be different from that along the [110], or OA, direction in Figure 5-34.

For a given direction in $\bar{k}$ space, where the isoenergy contours are spherical, the electrons are nearly free; the Sommerfeld distribution function (Equation 5-21) is followed and N(E) varies as $E^{1/2}$. However, as the wave vector approaches $\bar{k}_c$, the density becomes greater than that predicted by the Sommerfeld parabola and the Fermi level is no longer spherical. States begin to fill in other directions. Fewer and fewer states become available as the higher energy levels in the corners of the zone are filled. At the point where the zone is completely filled, N(E) equals zero. This relationship is illustrated in Figure 5-35 for a two-dimensional cubic lattice.

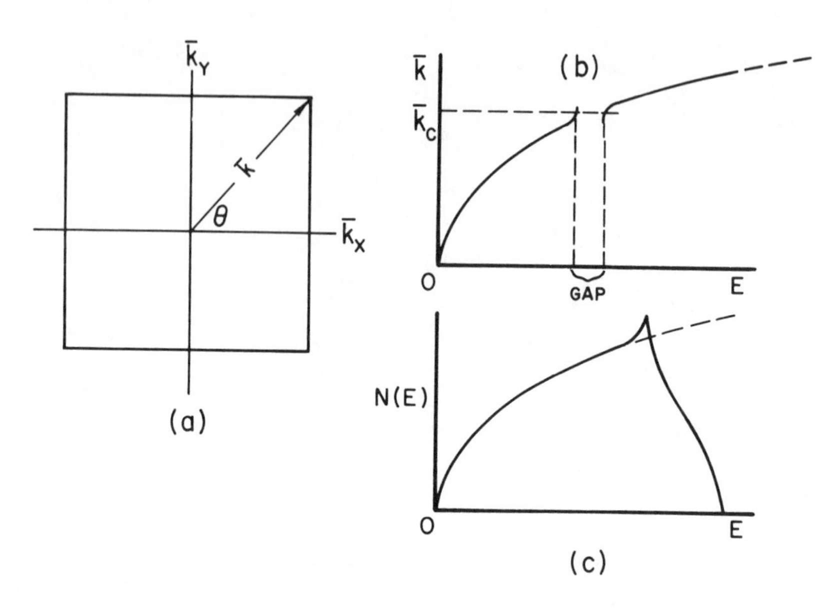

FIGURE 5-35. Relationships between band and zone theories. (a) Brillouin zone; (b) band theory; (c) curve for the density of states for a given zone.

The directional properties of the E vs. $\bar{k}$ relationship are important because they permit the prediction of properties as a function of crystalline direction within the zone. What appears to be a large energy gap in one direction may be spanned in another. This is shown in Figure 5-36. If this zone appeared to be filled in the $\bar{k}_x$ direction, the solid could be mistaken for a semiconductor or an insulator, if only the one direction were considered. The overlap of the states in the [110] direction with the second zone indicates that no effective gap exists; such a solid will demonstrate metallic behavior.

The solid under consideration could have had an effective gap if the curve for the [110] direction had shown a discontinuity at $4 < E < 5$ (arbitrary units). The energy gap between the first and second zones would then have consisted of about one energy unit. If this gap had been of the order of k_BT, the solid would have been classified as a semiconductor. Had it been much longer than k_BT, it would have been classed as an insulator.

The normal metals, such as copper, silver, or gold, each have one electron per atom; their zones thus are half-filled. At this electron concentration the isoenergy contours may be approximated to be spherical in $\bar{k}$ space; that is, their curves of density of states are considered to lie on the parabolic portion of the Sommerfeld curve. This is an approximation, since, in actuality, small "necks" protrude from the sphere to the hexagonal, octahedral planes of the first Brillouin zone of these FCC elements. The spherical approximation of the Fermi surface for such elements gives reasonable results and avoids the complications introduced by the necks. This approximation is applied to alloy phase formation in Chapter 10, Volume 3.

The variation of electron energy with the wave vector, as in Figure 5-36, has important consequences best visualized by means of the Brillouin zone. Since E varies with $\bar{k}$, N(E) also will be affected by this variation. The resulting anisotropy in N(E) will give corresponding variations in $n(E_F)$. (It will be recalled that $n(E_F)$ is the number of electrons per unit volume having $E \geq E_F$.) Since these are the electrons which enter into physical processes, variations in their number strongly affect those physical properties which take $n(E_F)$ into account.

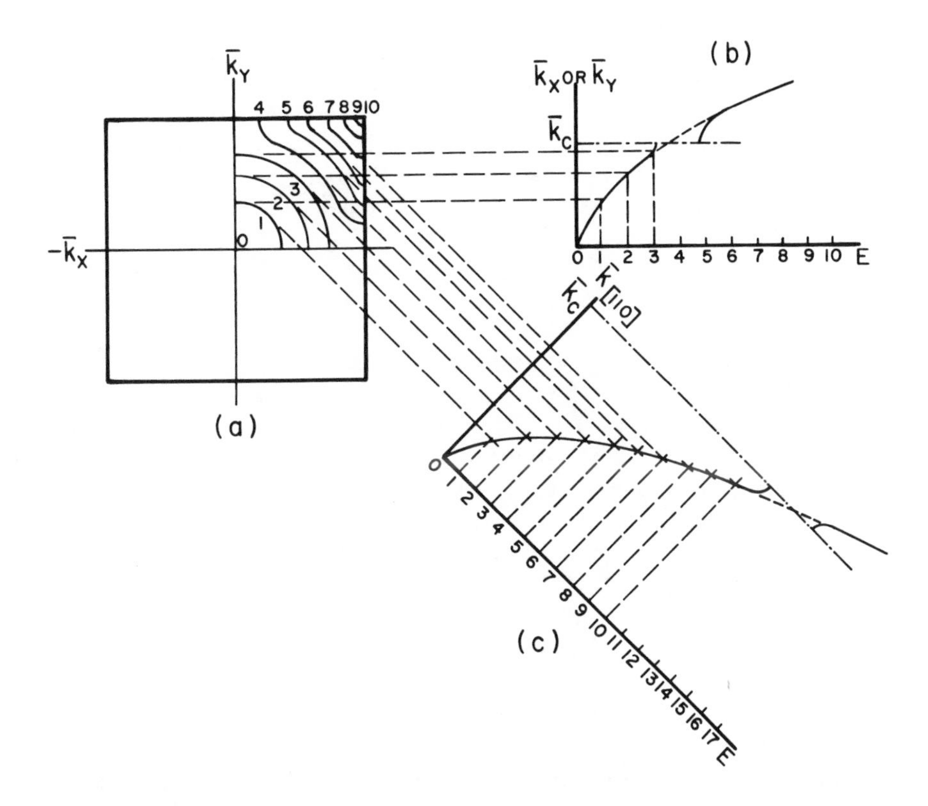

FIGURE 5-36. Variation of wave vector with direction in zone. (a) Isoenergy contours in the Brillouin zone; (b) variation of $\bar{k}$ with E in the $\bar{k}_x$ or $\bar{k}_y$ directions; (c) variation of $\bar{k}_{[110]}$ with E (diagonal direction). E and $\bar{k}$ in arbitrary units. (Method after Hume-Rothery, W., *Atomic Theory for Students of Metallurgy,* The Institute of Metals, London, 1952, 204.)

The E vs. $\bar{k}$ variations within a given Brillouin zone also help to explain the properties of alloys and alloy phases. This is especially true when the Brillouin zone is more than half filled. For example, if there are almost as many valence electrons as there are available states within a zone, it is to be expected that $n(E_F)$ will be quite small, in the absence of overlapping zones. In such a case, physical properties involving this factor would be significantly different from the situation in which $n(E_F)$ was comparatively large. This also is important in the anisotropy of mechanical properties of materials with HCP lattices.

Another consequence of the variation of electron energy with the wave vector within the zone results in a corresponding anisotropy in the effective mass of the electron. This follows from its behavior, previously noted in Section 5.8.2, as a function of $(d^2E/d\bar{k}^2)^{-1}$. This, too, has a profound influence upon the variation of many physical properties as a function of crystal direction.

The anisotropic effects observed in many crystals may be explained when both of these factors are taken into consideration. Such behavior may be illustrated by examining the electrical conductivity of metals as given by

$$\sigma = \frac{n(E_F)e^2\tau(E_F)}{m} \tag{5-64}$$

This relationship includes both factors under consideration. Differences in both $n(E_F)$ and m for various $\bar{k}$ can explain the observed anisotropy in the conductivity of single crystals.

The anisotropy of physical properties must be taken into account in applying engineering materials to obtain their optimum utilization. This is particularly true of wrought (textured) titanium and other HCP alloys. In such cases, the elastic constants, coefficients of thermal expansion and thermal conductivities vary considerably. Other materials, such as those used for hard and soft magnets and for semiconductor devices, are processed to have specific crystal orientations to achieve the most desirable properties.

5.9. PROBLEMS

1. Given that the electrical resistivity of copper is 1.71×10^{-6} ohm-cm at 25°C, calculate the mean free path, velocity, and relaxation time of an electron as predicted by the Drude-Lorentz theory.
2. Use the above information and Wiedemann-Franz ratio to calculate the thermal conductivity of copper. Compare your results with the data given in Table 5.6.
3. Discuss the Drude-Lorentz dilemma for multivalent metals.
4. Approximate the average energies of the valence electrons of Cu, Ag, and Au, if their Fermi energies are those given.
5. Make a graph of the Fermi-Dirac expression as a function of T for $(E-F_F) = \pm k_BT$ and for three selected values of k_BT.
6. Calculate the relaxation time for copper electrons using the data given in Question 1 and the Sommerfeld equation. Compare your results with those obtained above.
7. Calculate $C_{v,e}$ for copper. Given that the specific heat of copper is 0.092 cal/g/°C at 20°C, what percent of the internal energy is contained in the electrons?
8. Repeat Problem 1 using Equation 5-69 and compare results.
9. Calculate the value of γ for gold. How does it compare with that given in Table 5-5? What value would you suggest for the parameter A?
10. Given that the value of $(E_O - E_F)$ for nickel is 0.80 eV, calculate $C_{v,e}$ for Ni. What value would you suggest for the parameter A′?
11. Calculate the energy of a free electron in silver, where the closest approach of atoms is 2.888×10^{-8} cm.
12. What is the value for $\overline{k}_c$ for silver in the [110]?
13. Define the planes of the first and second Brillouin zones of the FCC lattice.
14. Define the planes of the first and second Brillouin zones of the BCC lattice.
15. Define the planes of the first and second Brillouin zones of the hexagonal lattice.
16. Calculate the c/a ratio for an ideal HCP lattice.
17. Determine the reciprocal lattice vectors for the HCP unit cell and express them in Equation 5-84b.
18. (a) Draw the curves for N(E) vs. E based upon zone theory for a univalent metal; (b) for a Divalent metal.
19. (a) Same as Problem 18, for an insulator; (b) For a semiconductor.

5.10. REFERENCES

1. **Seitz, F.,** *The Physics of Metals,* McGraw-Hill, New York, 1943.
2. **Kittel, C.,** *Introduction to Solid State Physics,* 3rd ed., John Wiley & Sons, New York, 1966.
3. **Hume-Rothery, W.,** *Atomic Theory for Students of Metallurgy,* Institute of Metals, 1952.
4. **Levy, R. A.,** *Principles of Solid State Physics,* Academic Press, New York, 1968.

5. **Hutchison, T. S., and Baird, D. C.,** *The Physics of Engineering Solids,* John Wiley & Sons, New York, 1968.
6. **Mott, N. F., and Jones, H.,** *The Theory of Properties of Metals and Alloys,* Dover, New York, 1958.
7. **Jones, H.,** *The Theory of Brillouin Zones and Electronic States in Crystals,* North-Holland, Amsterdam, 1960.
8. **Sproull, R. L.,** *Modern Physics,* John Wiley & Sons, New York, 1956.
9. **D'Abro, A.,** *The Rise of the New Physics,* Dover, New York, 1951.
10. **Richtmyer, F. K., Kennard, E. H., and Lauritsen, T.,** *Introduction to Modern Physics,* McGraw-Hill, New York, 1955.
11. **Stringer, J.,** *An Introduction to the Electron Theory of Solids,* Pergamon Press, Elmsford, N.Y., 1967.
12. **Cottrell, A. H.,** *Theoretical Structural Metallurgy,* St. Martin's Press, New York, 1957.
13. **Raimes, S.,** *The Wave Mechanics of Electrons in Metals,* North-Holland, Amsterdam, 1961.
14. **Lyman, T., Ed.,** *Metals Handbook,* Vol. 1, American Society for Metals, Metals Park, Ohio, 1961.
15. **Brillouin, L.,** *Wave Propagation in Periodic Structures,* Dover, New York, 1953.
16. **Martin, T. L., Jr. and Leonard, W. F.,** *Electrons and Crystals,* Brooks/Cole, Monterey, Ca., 1970.
17. **Barrett, C. S. and Massalski, T. B.,** *Structure of Metals,* McGraw-Hill, New York, 1966.

APPENDIX A

USEFUL PHYSICAL CONSTANTS

Constant	Symbol	Value: CGS	Value: SI
Electron rest mass	m	9.11×10^{-28} g	9.11×10^{-31} kg
Electron charge	e	4.80×10^{-10} esu	1.60×10^{-19} coul
Planck's constant	h	6.63×10^{-27} erg sec	6.63×10^{-34} J sec
Planck's constant/2π	$\hbar$	1.05×10^{-27} erg sec	1.05×10^{-34} J sec
Boltzmann's constant	k_B	1.38×10^{-16} erg/K	1.38×10^{-23} J/K
	k_B	8.63×10^{-5} eV/K	
Electron volt	eV	1.60×10^{-12} erg	1.60×10^{-19} J
Electron volt/molecule	eV/a	23.06 kcal/mol	
Gas constant	R	1.987 cal/(mol K)	8.31 J/(mol K)
Avogadro's number	N, N_A	6.02×10^{23}/mol	6.02×10^{26}/kmol
Atomic mass unit	amu	1.66×10^{-24} g	1.66×10^{-27} kg

APPENDIX B

CONVERSION OF UNITS

Unit	Symbol	Conversion operation	Resulting units
Electrical potential (volt)	V	V/300	statvolt
Electrical current (amp)	A	$A/(3 \times 10^9)$	statamp
Electric field (V/cm)	$\bar{E}$	$\bar{E}/300$	statvolt/cm
Conductivity [(Ω-cm)$^{-1}$]	σ	$\sigma \times 9 \times 10^{11}$	esu conductivity
Resistivity (Ω-cm)	ϱ	$\varrho/(9 \times 10^{11})$	esu resistivity
Mobility [cm^2/(V sec)]	μ	$\mu \times 300$	cm^2/statvolt sec
Current density (amp/cm^2)	j	$j \times 3 \times 10^9$	statamp/cm^2
Magnetic flux density (Weber/m^2)	T	$T \times 10^4$	gauss
Magnetic field strength (amp turns/m)	H	$H \times 4\pi \times 10^{-3}$	oersted
Thermal conductivity [cal/(cm sec°C]	$\varkappa$	$\varkappa \times 422$	Watt/(m°K)

INDEX

A

B

C

D

 Sommerfeld, see Sommerfeld distribution

E

F

G

J

K

L

M

N

Q

R

S

T

U

V

W

X

Y

Z